Antennas

Volume 1

General Principles

Antennas

Volume 1
General Principles

E Roubine and J C Bolomey

Translated by
Meg Sanders

HEMISPHERE PUBLISHING CORPORATION
A subsidiary of Harper & Row, Publishers, Inc.
Washington New York London

Revised and updated 1987

English edition first published 1987 by North Oxford Academic Publishers Ltd, a subsidiary of Kogan Page Ltd, 120 Pentonville Road, London N1 9JN

Published in the U.S.A. by
Hemisphere Publishing Corporation

Library of Congress Cataloging-in-Publication Data
Antennes. English.
Antennas.
Translation of: Antennes.
Includes bibliographies and indexes.
Contents: v. 1. General principles / E. Roubine and J. C. Bolomey—v. 2. Applications / S. Drabowitch and C. Ancona.
1. Antennas (Electronics) I. Roubine, E. (Elie) II. Title.
TK7871.6.A513 1987 621.38′028′3 87–12111
ISBN 0–89116–278–X (v. 1)
ISBN 0–89116–279–8 (v. 2)

Set by Eta Services (Typesetters) Ltd, Beccles, Suffolk
Printed and bound in Great Britain

Contents

Foreword

This work was inspired by lectures given by its authors at the École Supérieure d'Électricité. It is not intended as a treatise on antennas. The mere size of this volume clearly rules out any such exhaustive study, which would fill an encyclopaedia. To pare down this enormous body of knowledge has been the greatest challenge to the authors, who have constantly felt the conflict between the necessary task of reducing and editing their lectures, and the desire to present modern and sometimes innovatory technical knowledge in as complete a way as possible.

The development of antennas, from the time of the first radio link established by Hertz in his well known experiment of 1887, shows the scale of the subject.

Marconi entered the field in 1901, with the first transatlantic reception, and this tremendous achievement has tended to overshadow those of the pioneers. But the success of his venture was based on Hertz's discharger and the antenna planned by Popov for the detection of storms over several kilometres, a Branly coherer attached to a lightning conductor (1896).

This early period was also marked by considerable theoretical activity. Hertz developed the theory of the dipole, Max Abraham that of the very thin prolate spheroid. Poincaré considered the cylinder, using integral equations. Even the results of Sommerfeld's work on terrestrial propagation, which formed a basis for the development of powerful long-wave links, had an influence on such large installations as Croix d'Hins or Sainte Assise (1921).

The distant ionospheric propagation of HF signals was discovered by amateurs shortly after the First World War. The period between the wars was marked by an increased number of arrays and rhombic antennas, while in parallel, radiodiffusion MF broadcasting was developed. This was the period of pylon antennas. Finally, the ideas of Hertz were readopted, with the first attempts at microwave direct links. The first parabolic reflector theory was developed by Darbord in 1931, and the notion of electromagnetic detection was already becoming accepted.

The use of radar in the Second World War led to an increased role for microwave antennas, the importance of which has been confirmed with the passing of time. In this context, electromagnetism proper once again took on the major role that the empiricism of the inter-war period, as exemplified by the use of the pylon antenna, had somewhat ignored. The return to electromagnetic orthodoxy also allowed the contradictions of the apparently natural model of the equivalent line to be overcome, and the observed impedances to be confirmed. This led to a turning point in the theory of common antennas: those developed along the lines of the cylindrical dipole (Hallen, 1938), the bicone

(Schelunkoff, 1941), and the prolate spheroid (Stratton and Chu, 1941) soon became famous. With hindsight, the role of the particularly difficult problem of input impedance can be seen to have been exaggerated, but the fact remains that research carried out in this context contributed to a clearer understanding of the functioning of antennas.

The benefits of electromagnetic theory in the area of microwaves have proved far more durable. It renewed the theory of diffraction, and the attention devoted to it in universities all over the world is significant of its basic importance.

The contemporary period of antennas is, above all, marked by the related needs of radar, radio astronomy and space technologies, with renewed analysis, which is frequently highly sophisticated (arrays, for example), using new and improved means of numerical calculation.

This rapid survey is intended not only to show show ambitious a project it would be to attempt an exhaustive treatise on the subject of antennas, but also to explain the need for economy in this work.

The two volumes that make up this work are each divided into two parts:

Volume 1	Part 1: Theoretical introduction (Roubine)
	Part 2: General properties of antennas (Bolomey)
Volume 2	Part 1: Large antennas (Drabowitch)
	Part 2: Small antennas and plasma-embedded antennas (Ancona)

Part 1 is a summary of the results of classical electromagnetism on which the theory of antennas is based. The intention is not to reconsider this body of knowledge, which has given rise to many remarkable discoveries, but simply to illustrate the contribution of modern analysis in the simplification of calculation and the generalization of their results. A certain degree of abstraction is accepted by today's students.

Less is generally known on the subject of optics, which was unfortunately not included in most study courses until quite recently. For this reason, the basic elements are summarized in Part 1, which ends with some of the concepts of electromagnetic diffraction and a study of various of its techniques.

Part 2 considers problems of fields and radiated power, gain, impedance and coupling, etc., common to all antennas the dimensions of which are measured in wavelengths. The content of this section is standard, but fundamental, and has been expressed from a modern viewpoint. A considerable amount of space has been devoted to the problem of reception and to numerical methods. These have had a significant effect on the subject, and justify the reduction in the consideration of theoretical developments in impedance to the minimum. The major concern has been to produce a text that can be used by engineers, without requiring an in-depth study of Part 1. The presentation of experimental techniques clearly underlies this preoccupation.

The second volume is deliberately devoted to applications, and is roughly divided between large and small antennas, based on the themes most representative of the 'state of the art', and which provide plenty of scope for instruction. The authors, practised engineers who have contributed to the development of their particular fields, are also lecturers, hence their interest in an integrated presentation that will be of practical use to future workers.

In Vol. 2, Part 1 basically concerns large (in terms of wavelength) antennas, such as the many ground-based antennas. This is a large subject in itself but, rather than compiling a catalogue of antennas, the author has concentrated on these methods developed in the most important recent studies and which are based on real models. A desire for greater coherence has led him to concentrate on revolving structures with rotational symmetry in which the concept of a hybrid mode allows primary sources, corrugated horns, diffraction at the focus and at infinity in reflectors of revolution to be treated with similar formulae.

The same preoccupations are evident in Part 2, which concerns a subject of no small importance, despite its title and the lack of attention devoted to it in education, but in which the antennas under consideration amply show its significance. Scientific advance in problems in this area is frankly difficult, yet technical developments, particularly in space links, are making the way forward clearer all the time. The study of antennas in an ionized environment is a typical example, and one with which the author is very familiar, having contributed as a physicist.

It is to be hoped that, despite the reservations expressed above, this work forms a coherent unit. Nonetheless, the authors were given considerable latitude, which may have resulted in an occasional lack of consistency in notation or some repetition, despite attempts at coordination.

My pleasure in presenting the contributions of my co-authors is due, in part, to the facts that all three have at some time been my students and—although the one is not necessarily due to the other—that all are respected experts in their field.

Finally, I would like to express my gratitude to the École Supérieure d'Électricité, Thomson-CSF and the Société Technique d'Application et de Recherche Electronique, whose material aid greatly assisted in the publishing of this work.

E Roubine

Part 1

Theoretical introduction

Chapter 1

Electromagnetic field of antennas

A: Purpose of Part 1

The first part of this work is intended as a summary of the elements of theoretical electromagnetism necessary to the study of the physical bases of antenna function. Despite the fact that these elements are to be found in many other texts, it was judged useful to present the most important results here, even if only to clarify the notation used. References are provided for derivations of a purely technical mathematical nature, which have not been included here.

It was felt that there could be considerable interest in presenting such a classic subject from a more modern viewpoint than is customary. The reader will notice the economy of method provided by the theory of distribution: almost immediate derivation of boundary conditions and of Huygens' theorem, direct expression of radiated fields without the use of potentials, simple handling of convolutions, etc.

Furthermore, the use of the complex combination $E + jH$ of the electric and magnetic fields, suggested by their correlative properties, was a deliberate choice. The idea is an old one (Silberstein, 1907), but seems to have been used only tentatively by other authors. The reason for this is, doubtless, the ambiguity that can be created by the complex representation of the fields themselves, with sinusoidal time dependence. The use of bicomplex algebra with four units allows the problem to be overcome. The compactness of expression which results is striking (Appendix 1).

At very high frequency traditional geometric optics is an approximation which is very useful in problems of radiation and diffraction. Certain basic results have been summarized here, for two reasons. The first is that optics has unfortunately been absent from most university electrical engineering courses for some years, and many students know little or nothing about it. The second reason is that it has given rise to an ingenious extension which in recent years has become standard for the solution of diffraction problems (the geometric theory of diffraction).

Diffraction is naturally involved in the study of antennas (reflectors, apertures, and reception antennas). This is why theory is discussed at the end of Part 1.

B: Maxwell's equations—physical viewpoint

1: Maxwell's equations

The electrical properties of any medium are determined by charges, currents and an electromagnetic (e.m.) field. If the charges possess, at each point $x = (x_1, x_2, x_3)$ and at each instant t the densities $\rho(x, t)$ (in C/m^3) and $\vec{\mathscr{J}}(x, t)$ (in A/m^2), these densities are related, as in fluid mechanics, by the conservation or continuity equation:

$$\operatorname{div} \vec{\mathscr{J}} + \partial_t \rho = 0 \tag{1}$$

The e.m. field is made up of four vectorial fields:

an electrical field and an electric induction: $\vec{\mathscr{E}}(x, t)$ (in V/m) and $\vec{\mathscr{D}}(x, t)$ (in C/m);

a magnetic field and a magnetic induction: $\vec{\mathscr{H}}(x, t)$ (in A/m) and $\vec{\mathscr{B}}(x, t)$ (in T [teslas]).

Maxwell's equations are the system of partial differential equations that describe the local interactions:

$$\operatorname{curl} \vec{\mathscr{E}} + \partial_t \vec{\mathscr{B}} = 0 \quad \text{(induction law)} \tag{2}$$

$$\operatorname{curl} \vec{\mathscr{H}} - \partial_t \vec{\mathscr{D}} = \vec{\mathscr{J}} \quad \text{(Ampere's law)} \tag{3}$$

$$\operatorname{div} \vec{\mathscr{D}} = \rho \quad \text{(Gauss's electric law)} \tag{4}$$

$$\operatorname{div} \vec{\mathscr{B}} = 0 \quad \text{(Gauss's magnetic law)} \tag{5}$$

Gauss's two laws are immediate consequences of (1), (2) and (3). For the present, only those areas in which the fields are twice continuously differentiable relative to the coordinates and time will be considered.

2: Simple (or perfect) media

These are characterized by two constants ε and μ such that the following relationships exist:

$$\vec{\mathscr{D}} = \varepsilon \vec{\mathscr{E}} \tag{6}$$

$$\vec{\mathscr{B}} = \mu \vec{\mathscr{H}} \tag{7}$$

The media are linear, isotropic and homogeneous. ε is the absolute permittivity (or absolute dielectric constant) and is expressed in F/m. μ is the absolute permeability, expressed in H/m.

The simplest example (and the most important in this context) is that of the vacuum: $\varepsilon = \varepsilon_0$, $\mu = \mu_0$, where the constants ε_0 and μ_0 are fixed by the system of units. In the so-called coordinated systems:

$$\varepsilon_0 \mu_0 c^2 = 1 \tag{8}$$

where c has the dimensions of velocity and is numerically equal to that of light *in vacuo*. The approximation $c = 3 \times 10^8$ m/s is quite sufficient here. So in the rationalized MKSA system:

$$\varepsilon_0 = \frac{1}{36\pi} 10^{-9} \quad \text{F/m}$$
$$\mu_0 = 4\pi\, 10^{-7} \quad \text{H/m} \tag{9}$$

With a more precise value ($c = 2.9979 \ldots \times 10^8$) one could, for example, adjust μ_0 versus ε_0.

Simple media are defined by a relative permittivity ε_r and a relative permeability μ_r, such that:

$$\varepsilon = \varepsilon_0 \varepsilon_r$$
$$\mu = \mu_0 \mu_r \tag{10}$$

The relative permeability and permittivity, both dimensionless, are given in tables.

3: Propagation of the electromagnetic field

Apart from the charges and currents, in simple media we have:

$$\begin{aligned} \text{curl}\, \vec{\mathscr{E}} + \mu \partial_t \vec{\mathscr{H}} &= 0 \\ \text{curl}\, \vec{\mathscr{H}} - \varepsilon \partial_t \vec{\mathscr{E}} &= 0 \\ \text{div}\, \vec{\mathscr{E}} &= 0 \\ \text{div}\, \vec{\mathscr{H}} &= 0 \end{aligned} \tag{11}$$

By introducing the operator $\Delta = \text{grad div} - \text{curl curl}$:

$$\begin{aligned} \Delta \vec{\mathscr{E}} - \varepsilon\mu \partial_{tt}^2 \vec{\mathscr{E}} &= 0 \\ \Delta \vec{\mathscr{H}} - \varepsilon\mu \partial_{tt}^2 \vec{\mathscr{H}} &= 0 \end{aligned} \tag{12}$$

The fields $\vec{\mathscr{E}}$ and $\vec{\mathscr{H}}$ obey the vectorial wave equation (Dalembert's equation). They propagate by waves with velocity:

$$v = \frac{1}{\sqrt{\varepsilon\mu}} = \frac{c}{\sqrt{\varepsilon_r \mu_r}} \tag{13}$$

In particular, they propagate in a vacuum with velocity c.

This first, major consequence of Maxwell's equations is the origin of Maxwell's electromagnetic theory of light (the electric vector being the light vector).

It was Hertz who, 20 years later (1887), confirmed the propagation of the e.m. field experimentally, and was thereafter considered to be the 'inventor' of radio. His experiments constituted the first transmissions of the type later to be called Hertzian. Hertz's discharger and resonator were the first antennas.

Note that that relationship (13) defines the refractive index of the medium:

$$n = \frac{c}{v} = \sqrt{\varepsilon_r \mu_r} \tag{14}$$

This permits design with lenses of normal ($\mu_r = 1$) or artificial dialectric.

4: Conductive media

(a) Conductivity

Conductive media obey Ohm's law:

$$\vec{\mathscr{J}} = \sigma \vec{\mathscr{E}} \tag{15}$$

Conductivity σ is expressed in Ω^{-1}/m or ℧/m (mhos/m). This is the reciprocal of resistivity (expressed in Ω m). In pure metals (except for mercury) it is of the order of 10^7 (Ag: 6.1×10^7; Cu: 5.8×10^7).

It is easy to verify that the relaxation relationship applies for internal charges:

$$\rho(x, t) = \rho(x, 0)e^{-t/\tau}, \quad \tau = \frac{\varepsilon}{\sigma} \tag{16}$$

where the relaxation time τ is extremely low in good conductors. The charges are carried to the surface almost instantaneously.

(b) Perfect conductors

The above suggests the idealized model of infinite conductivity. This simplifies the calculations for metallic conductors to a considerable degree, and the results constitute an approximation which is largely adequate for most applications.

(c) Applied current

It is sometimes useful, in the study of antennas, to distinguish between 'ohmic' current (15) and 'external' current $\vec{\mathscr{J}}_a$ considered as a source applied to the antenna. In regions where the two coexist, there will be a total current $\sigma\vec{\mathscr{E}} + \vec{\mathscr{J}}_a$, which will constitute the second term of (3).

5: Anisotropic media

These are used in Vol. 2, Part 2. The inductions are only related linearly to the corresponding fields. In a cartesian coordinate system (which could be assumed to be rectangular), the following relationships exist between components:

$$\mathscr{D}_i = \sum_j \varepsilon_{ij} \mathscr{E}_j \tag{17}$$

$$\mathscr{B}_i = \sum_j \mu_{ij} \mathscr{H}_i \tag{18}$$

or in matrix form, for example for the first:

$$\begin{pmatrix} \mathscr{D}_1 \\ \mathscr{D}_2 \\ \mathscr{D}_3 \end{pmatrix} = \begin{pmatrix} \varepsilon_{11} & \varepsilon_{12} & \varepsilon_{13} \\ \varepsilon_{21} & \varepsilon_{22} & \varepsilon_{23} \\ \varepsilon_{31} & \varepsilon_{32} & \varepsilon_{33} \end{pmatrix} \begin{pmatrix} \mathscr{E}_1 \\ \mathscr{E}_2 \\ \mathscr{E}_3 \end{pmatrix} \tag{19}$$

In a homogeneous medium the coefficients are constants, but they depend on the coordinate system chosen (e_1, e_2, e_3). An intrinsic significance is often given to the above relationships by assigning a tensorial permittivity and permeability to the medium:

$$\sum_{i,j} \varepsilon_{ij}\vec{e}_i \otimes \vec{\varepsilon}_j \quad \text{and} \quad \sum_{i,j} \mu_{ij}\vec{e}_i \otimes \vec{e}_j$$

C: Sinusoidal time dependence

1: Sinusoidal fields

From now on it will be assumed that all the variables vary sinusoidally in time with the same angular frequency $\omega = 2\pi f$. All the scalars, notably the field components, in any Cartesian coordinate system, are in the form:

$$\mathscr{E}_k(x, t) = E_{k,0}(x) \cos [\omega t + \varphi_k(x)] = A_k(x) \cos \omega t - B_k(x) \sin \omega t \quad (k = 1, 2, 3) \tag{20}$$

With the additional vectors $\vec{A} = (A_1, A_2, A_3)$ and $\vec{B} = (B_1, B_2, B_3)$ the following can be expressed:

$$\vec{\mathscr{E}}(x, t) = \vec{A}(x) \cos \omega t - \vec{B}(x) \sin \omega t \tag{21}$$

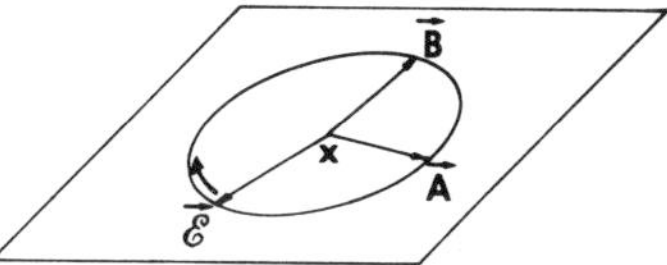

Fig. 1. *Elliptical polarization.*

One recognizes a parametric representation of an ellipse with the two conjugate radii $\vec{A}$ and $\vec{B}$ (which are assumed to be non-parallel). The ellipse, centred on $x = (x_1, x_2, x_3)$, is scanned at frequency f by the vector $\vec{\mathscr{E}}$.

If the additional vectors were parallel, (21) would represent a rectilinear oscillation on the common axis. This is the particular case with a completely flat ellipse.

Another interesting example is the circular case ($\vec{A} \perp \vec{B}$ and $A = B$).

The sinusoidal fields are therefore always elliptically polarized (in particular, rectilinearly).

2: Complex fields

Using the well known, complex representation of components

$$E_k(x) = E_{k,0}(x)\, e^{i\varphi_k(x)} = A_k(x) + iB_k(x) \tag{22}$$

the calculations are considerably simplified. In the real field $\vec{\mathscr{E}}(x, t)$, a function of x and t, a complex field is associated:

$$\vec{E}(x) = (E_1(x), E_2(x), E_3(x)) = \vec{A}(x) + i\vec{B}(x) \tag{23}$$

which depends only on x. Inversely, the three relationships

$$\mathscr{E}_k(x, t) = \mathscr{R}e\,\{E_k(x)\,e^{i\omega t}\}, \quad k = 1, 2, 3 \tag{24}$$

can be symbolically summarized by:

$$\vec{\mathscr{E}}_k(x, t) = \mathscr{R}e\,\{\vec{E}(x)\,e^{i\omega t}\} \tag{25}$$

which is, in fact, an extremely useful compact rule for the interpretation of results for any calculation on the complex fields.

Exercises

- Interpret $\lambda\vec{E}(x)$ (λ: constant complex scalar) and $\vec{E}^*(x)$.
- Interpret $\vec{E}e^{-i\vec{k}\cdot\vec{x}}$ ($\vec{E} = \vec{A} + i\vec{B}$: elliptical vector independent of x; $\vec{k}$: constant real vector (propagation by homogeneous plane waves in the direction of the propagation vector $\vec{k}$).

Notes

- The complex representation

$$\vec{E}_t(x) = \vec{E}(x)\,e^{i\omega t} \tag{26}$$

explicitly containing time is frequently used by optical scientists. This gives directly:

$$\vec{\mathscr{E}}(x, t) = \mathscr{R}e\,\{\vec{E}_t(x)\} = \frac{1}{2}\,\vec{E}_t(x) + \frac{1}{2}\,\vec{E}_t^*(x) \tag{27}$$

Electrical engineers do not use the exponential, which disappears in homogeneous linear relationships (such as Maxwell's equations). This is why it must be restored in (25).

- The sinusoidal signals correspond to an actual situation, in which, in the modulation process, there are narrow-band signals around a carrier ω. The general case can be treated using a Fourier Transform with respect to time,

$$\mathscr{F}\{\vec{\mathscr{E}}(x, t)\} = \vec{\mathscr{E}}^{\wedge}(x, \omega) \tag{28}$$

which for each ω reduces to the preceding formula. The complex representation therefore has the additional advantage of being a calculation intermediary for the Fourier syntheses. More precisely, for sinusoidal signals, the Fourier transforms can be expressed as:

$$\frac{1}{2\pi}\,\mathscr{F}\{\vec{E}_t(x)\} = \vec{E}(x)\delta_{(\omega)} \tag{29}$$

$$\frac{1}{\pi}\,\mathscr{F}\{\vec{\mathscr{E}}(x, t)\} = \vec{E}(x)\,\delta_{(\omega)} + \vec{E}^*(x)\,\delta_{(-\omega)} \tag{30}$$

($\delta_{(\omega)}$ being the Dirac function at frequency ω). All these considerations are based on any fixed point x. The Fourier transforms in these coordinates, which play a very important role in the theory of antennas, are discussed below.

• Many authors (particularly Americans) prefer the temporal factor $e^{-i\omega t}$, although this does not change (25). The linkage of notation is easy: it is possible to pass from one set to the other by conjugation of all the complex values. The main interest of this convention is that plane waves are characterized by a propagation factor $e^{ik\cdot x}$ (see second exercise): the propagation takes place in the direction of increasing coordinate.

3: Rules for calculation

Complex representation comes down to providing each point x with a tangent vectorial space, identifiable as C^3. The definitions and operations of normal vector calculus are transposed, in a coherent way, by means of the components. The only problems that may arise concern the physical interpretation of the results. The linear calculation is straightforward:

$$\mathscr{E}\cdot\vec{\alpha}, \vec{\mathscr{E}} \wedge \vec{\alpha}(\vec{\alpha}\text{: constant vector}), \text{ grad } \mathscr{V}, \text{ div } \vec{\mathscr{E}} \text{ curl } \vec{\mathscr{E}}$$

have as their respective images:

$$\vec{E}\cdot\vec{\alpha}, \vec{E} \wedge \vec{\alpha}, \text{ grad } V, \text{ div } \vec{E}, \text{ curl } \vec{E}$$

The temporal derivative $\partial_t\vec{\mathscr{E}}$ has as its image $i\omega\vec{E}$.

The only problems arise in the interpretation of the complex scalar and vector products. This will be further considered at a later stage.

4: Maxwell's equations

Using the above rules, the following can be established immediately:

$$\begin{aligned} \text{curl } \vec{E} + i\omega\vec{B} &= 0 \\ \text{curl } \vec{H} - i\omega\vec{D} &= \vec{J} \\ \text{div } \vec{D} &= \rho \\ \text{div } \vec{B} &= 0 \end{aligned} \tag{31}$$

with

$$\text{div } J + i\omega\rho = 0 \tag{32}$$

In this case, $\rho(x)$ is clearly the complex representation of the sinusoidal density $\rho(x, t)$.

In a simple medium:

$$\begin{aligned} \text{curl } \vec{E} + i\omega\mu\vec{H} &= 0 \\ \text{curl } \vec{H} - i\omega\varepsilon\vec{E} &= \vec{J} \\ \text{div } \vec{E} &= \frac{\rho}{\varepsilon} \\ \text{div } \vec{H} &= 0 \end{aligned} \tag{33}$$

From this it is possible to deduce, apart from the charge and currents, the vectorial Helmholtz equations

$$\begin{aligned} \Delta\vec{E} + k^2\vec{E} &= 0 \\ \Delta\vec{H} + k^2\vec{H} &= 0 \end{aligned} \tag{34}$$

obeyed by the field, where

$$k^2 = \omega^2\varepsilon\mu \tag{35}$$

or, using (13),

$$k = \frac{\omega}{v} \tag{36}$$

The Helmholtz equation is fundamental, and is the monochromatic wave equation (also known as the reduced wave equation).

k will henceforth play the essential role of phase constant. In a vacuum, it will be necessary to impose $k_0 = \dfrac{\omega}{c}$. In a conductive medium the second of the equations in (33) will be reduced to:

$$\text{curl}\,\vec{H} - (i\omega\varepsilon + \sigma)\vec{E} = 0 \tag{37}$$

from which can be deduced div $\vec{E} = 0$ and consequently $\rho = 0$.

It is helpful to introduce a complex permittivity:

$$\varepsilon_c = \varepsilon - i\,\frac{\sigma}{\omega} = \varepsilon(1 - i\tan\delta) = \frac{\varepsilon}{\cos\delta}\,e^{-i\delta} \tag{38}$$

with:

$$\tan\delta = \frac{\sigma}{\omega\varepsilon} \tag{39}$$

which comes back to the formula for simple media without sources.

The following system is found:

$$\begin{aligned} \text{curl}\,\vec{E} + i\omega\mu\vec{H} &= 0 \\ \text{curl}\,\vec{H} - i\omega\varepsilon_c\vec{E} &= 0 \\ \text{div}\,\vec{E} &= 0 \\ \text{div}\,\vec{H} &= 0 \end{aligned} \tag{40}$$

Notes

• The imaginary part of ε_c corresponds to Joule losses. In the usual dielectrics:

$$\sigma \ll \omega\varepsilon \quad \text{or} \quad \delta \ll 1 \tag{41}$$

Tan δ is therefore simply the loss tangent. (Example: for polythene,

$\tan\delta \simeq 10^{-4}$.) In good conductors:

$$\sigma \gg \omega\varepsilon \quad \text{or} \quad \delta \simeq \frac{\pi}{2} \tag{42}$$

Tan δ is — to the nearest right angle — the ratio of the conduction current to the displacement current. What distinguishes conductive media from insulating media, in macroscopic theory, is the predominance of one of the currents over the other. As shown in (39), it may happen that a medium is a dielectric at high frequencies and a conductor at low frequencies (for example, soil). Finally, in similarity techniques (reduced or enlarged models) in which ε can be considered as varying only slightly with frequency, the same values of δ will be used, taking care that the conductivity (particularly of the soil) is modified by the ratio of frequencies.

• By analogy with (35), it is of interest to consider propagation in conductors using a complex constant k_c such that:

$$k_c^2 = \omega^2 \varepsilon_c \mu \tag{43}$$

or

$$k_c = \frac{k}{\sqrt{\cos\delta}} e^{-i(\delta/2)} \tag{44}$$

In a good conductor, $\frac{\pi}{2} - \delta$ (δ small) is equivalent, using (39), to $\omega\varepsilon/\sigma$. Therefore:

$$k_c \simeq \frac{k}{\sqrt{\frac{\omega\varepsilon}{\sigma}}} e^{-i(\pi/4)} = \sqrt{\sigma\mu\omega}\, e^{-i(\pi/4)} \tag{45}$$

The propagation factor $e^{-ik_c z}$ in the direction of a coordinate z displays a linear attentuation:

$$\mathscr{R}\text{e}\{-ik_c\} = \frac{1}{p} = \sqrt{\frac{\sigma\mu\omega}{2}} = \sqrt{\pi\sigma\mu f} \qquad \text{népers/m} \tag{46}$$

p is the penetration in the conductor (also referred to, figuratively, as the skin depth). p is smaller when the frequency f or the conductivity σ is high. The limiting case of the perfect conductor ($\sigma = \infty$) implies $p = 0$. In a medium of this type, all the e.m. values are zero.

5: Extension to the products of complex vectors

(a) Products of complex fields

The space C^3 attached to any x may be formally given a scalar pseudo-product

$$\vec{E}\cdot\vec{F} = E_1F_1 + E_2F_2 + E_3F_3 \tag{47}$$

and a vectorial pseudo-product

$$\vec{E} \wedge \vec{F} = (E_2F_3 - F_3F_2, E_3F_1 - E_1F_3, E_1F_2 - E_2F_1) \tag{48}$$

A hermitian product may also be introduced:

$$(\vec{E}, \vec{F}) = \vec{E} \cdot \vec{F}^* \tag{49}$$

(where, in (47), the direction of scanning of the ellipse (F) is simply changed). In particular:

$$\vec{E} \cdot \vec{E}^* = (\vec{E}, \vec{E}) = \sum_k |E_k|^2 = E'^2 \tag{50}$$

is real and positive, and defines the norm of the field, which will be denoted E', to simplify matters, and which will be used consistently below (in fact $E' = \|\vec{E}\|$).

Notes

- For elliptical fields, the norm plays the role of the amplitude of rectilinear fields. Taking account of (22), the following can be noted:

$$E'^2 = |\vec{A}|^2 + |\vec{B}|^2 \tag{51}$$

Geometrically E' would be the radius of the Monge circle for the ellipse of polarization.

- A field of norm unity will be referred to as normalized. $\vec{E}/E'$ is normalized.

(b) Geometric interpretation

Complex fields are calculational intermediaries which are not always suitable for a simple transposition in real fields. For this reason a condition of orthogonality such as:

$$\vec{E} \cdot \vec{F} = 0 \quad \text{or} \quad \vec{E} \cdot \vec{F}^* = 0 \tag{52}$$

does not in any way imply, contrary to what has sometimes been written, orthogonality of the fields $\vec{\mathscr{E}}$ and $\vec{\mathscr{F}}$ at all times.

The following problem, which is encountered in the study of receiving antennas, can be looked upon as an exercise.

Exercise

Consider two polarization ellipses (E) and (F), which are coplanar and referred to their respective axes. A mutual polarization factor is defined:

$$\rho = \frac{\vec{E} \cdot \vec{F}}{E'F'} \tag{53}$$

such that $|\rho| \leqslant 1$. Show that:

(a) $|\rho| = 1$, when the two ellipses are homothetic and scanned in opposite directions.

(b) $\rho = 0$, when the two ellipses are similar, crossed (ie with orthogonal major axes) and scanned in the same direction. Study the specific examples of rectilinear and circular polarization.

Application (receiving antennas): Let $\vec{U}$ and $\vec{V}$ be two fields, coplanar in x, normalized and orthogonal, in the sense given in (b):

$$\begin{aligned} U' = V' = 1 \\ \vec{U} \cdot \vec{V} = 0 \end{aligned} \tag{54}$$

Any coplanar field $\vec{E}$ can be resolved according to the base $(\vec{U}, \vec{V})$:

$$\vec{E} = E_u \vec{U} + E_v \vec{V} \tag{55}$$

$$E_u = \vec{E} \cdot \vec{U}, \qquad E_v = \vec{E} \cdot \vec{V} \tag{56}$$

(c) Energy interpretation

The following calculation will be familiar to electrical engineers. Consider a dipole circuit with the following current flow:

$$i(t) = I_0 \cos(\omega t + \varphi) \tag{57}$$

and a voltage

$$v(t) = V_0 \cos(\omega t + \psi) \tag{58}$$

By expressing the instantaneous power in the form

$$\begin{aligned} w(t) &= v(t)i(t) \\ &= \frac{1}{2} V_0 I_0 \{\cos(\psi - \varphi)[1 + \cos 2(\omega t + \varphi)] \\ &\quad - \sin(\psi - \varphi) \sin 2(\omega t + \varphi\} \\ &= W'[1 + \cos 2(\omega t + \varphi)] + W'' \sin 2(\omega t + \varphi) \end{aligned} \tag{59}$$

a power flow in the same direction as an average (active power):

$$W' = \frac{1}{2} V_0 I_0 \cos(\psi - \varphi) \tag{60}$$

is superimposed on a power produced alternately in one direction, then the other, with an average value of zero. The intensity of these oscillations, known as reactives, is the reactive power:

$$W'' = \frac{1}{2} V_0 I_0 \sin(\psi - \varphi) \tag{61}$$

A complex power is defined by the conventional association:

$$W = W' + iW'' = \frac{1}{2} VI^* \tag{62}$$

where $V = V_0\, e^{i\psi}$ and $I = I_0\, e^{i\varphi}$ are the complex potential and current.

Let us recall that $W'' = 0$ is a resonance condition.

Note

The typical notation (62) is not a consequence of the complex representation, as summarized at the start. It results from a new convention. This point is not always made in literature on the subject.

From the approach mentioned above, an average localized energy can be envisaged for any simple dielectric of limited domain Ω:

$$T = \frac{1}{4}\int_{\Omega} (\varepsilon E'^2 + \mu H'^2)\, dx = T_E + T_H \tag{63}$$

and likewise in any conductive part Ω' the active Joule power:

$$W_j = \frac{1}{2}\int_{\Omega'} E'^2\, dx = \frac{1}{2}\int_{\Omega'} \frac{J'^2}{\sigma}\, dx \tag{64}$$

Finally, applied sources $\vec{J}_a$ will deliver the complex supply power:

$$W_a = W'_a + iW''_a = -\frac{1}{2}\int E \cdot \vec{J}_a^*\, dx \tag{65}$$

Notes

- In any medium, the following is adopted:

$$T_E = \frac{1}{4}\int \vec{D} \cdot \vec{E}^*\, dx \tag{66}$$

$$T_H = \frac{1}{4}\int \vec{B} \cdot \vec{H}^*\, dx \tag{67}$$

- The average energies T_E and T_H have an important role in some theoretical problems. Their finiteness in the area around an edge or wedge of a conductor is a uniqueness condition (Meixner's condition). It expresses the fact that the fields are square-integrable in such an area.

(d) Application to Poynting's theorem

Consider a restricted domain Ω, limited by a closed surface S containing a simple medium (ε, μ), conductive parts Ω', and applied sources. From Maxwell's equations:

$$\begin{aligned} -E \cdot \vec{J}_a^* &= -\vec{E}\ \text{curl}\ \vec{H}^* - i\omega\varepsilon E'^2 + \sigma E'^2 \\ &= \vec{H}^* \cdot \text{curl}\ \vec{E} - \vec{E} \cdot \text{curl}\ \vec{H}^* + i\omega\mu H'^2 - i\omega\varepsilon E'^2 + \sigma E'^2 \\ &= \text{div}\,(\vec{E} \wedge \vec{H}^*) + i\omega(\mu H'^2 - \varepsilon E'^2) + \sigma E'^2 \end{aligned}$$

Integrating with respect to Ω:

$$W_a = W_r + W_j + 2i\omega(T_H - T_E) \tag{68}$$

where

$$W_r = W'_r + iW''_r = \int_S P_n\, dS \tag{69}$$

Flux leaving S from the complex Poynting vector,

$$\vec{P} = \vec{P}' + i\vec{P}'' = \frac{1}{2}\vec{E} \wedge \vec{H}^* \tag{70}$$

has the significance of a complex power transmitted by radiation through S.

The energy balance (68) constitutes Poynting's theorem. By separating the real and imaginary parts:

$$W'_a = W'_r + W_j \tag{71}$$

$$W''_a = W''_r + 2\omega(T_H - T_E) \tag{72}$$

it can be stated that:

• The active supply power heats the conductors contained in Ω and yields an active power radiated through S to the outside. The result (71) is that this power is independent of any surface that contains the sources and the conductors. This is why it is possible to consider the active power radiated by a transmitting antenna independently of the closed surface used to calculate it.

• The reactive supply power, interpreted less strictly, is broken down into 'radiated' reactive power and a term proportional to the average excess magnetic energy located in Ω above the average electrical energy.

W''_r depends on S but it will be seen that it tends to cancel itself out when S approaches infinity. The usefulness of complex powers resides in their relationship with the impedances (eg the input or output impedance of an antenna).

(e) Application in anisotropic media

Consider a medium, without losses, which is only dielectrically anisotropic. For sinusoidal fields, (17) still applies with all the complex terms and, if there are no sources, the following can be written:

$$\text{curl}\,\vec{E} = -i\omega\mu\vec{H}$$

but:

$$\text{curl}\,\vec{H} = i\omega\vec{D} \tag{73}$$

will mean

$$(\text{curl}\,\vec{H})_k = i\omega \sum_l \varepsilon_{kl} E_l \tag{74}$$

then, as above:

$$\text{div}\,\vec{P} = \vec{H}^* \cdot \text{curl}\,\vec{E} - \vec{E} \cdot \text{curl}\,\vec{H}^* = -i\omega\mu H'^2 + i\omega \sum_{k,l} \varepsilon^*_{kl} E_k E^*_l \tag{75}$$

But according to the hypotheses, $W'_r = 0$ for any closed surface, so, of necessity $\vec{P}' = 0$ ($\vec{P}'$: real part of Poynting's vector). The second member of (75) is then purely imaginary. The last sum is real, so:

$$\sum_{k,l} \varepsilon^*_{kl} E_k E^*_l = \left(\sum_{k,l} \varepsilon^*_{kl} E_k E^*_l\right)^*$$

and as a result:

$$\varepsilon_{kl} = \varepsilon_{lk}^{*} \tag{76}$$

The permittivity matrix is hermitian, and can therefore be diagonalized according to a real orthogonal coordinate set, and written as below:

$$D_k = \varepsilon_{kk} E_k \tag{77}$$

(f) Application to the reciprocity theorem

Consider two distributions of sources, $\vec{J}_1$ and $\vec{J}_2$ in an unbounded simple medium (ε, μ), creating the respective fields ($\vec{E}_1, \vec{H}_1$ and $\vec{E}_2, \vec{H}_2$). Then:

$$\begin{aligned} \operatorname{div}(\vec{H}_2 \wedge \vec{E}_1) &= \vec{E}_1 \cdot \operatorname{curl} \vec{H}_2 - \vec{H}_2 \cdot \operatorname{curl} \vec{E}_1 \\ &= \vec{E}_1 \cdot (\vec{J}_2 + i\omega\varepsilon\vec{E}_2) + i\omega\mu\vec{H}_1 \cdot \vec{H}_2 \\ &= \vec{E}_1 \cdot \vec{J}_2 + i\omega(\varepsilon\vec{E}_1 \cdot \vec{E}_2 + \mu\vec{H}_1 \cdot \vec{H}_2) \end{aligned} \tag{78}$$

By changing the indices:

$$\operatorname{div}(\vec{H}_2 \wedge \vec{E}_1 - \vec{H}_1 \wedge \vec{E}_2) = \vec{E}_1 \cdot \vec{J}_2 - \vec{E}_2 \cdot \vec{J}_1 \tag{79}$$

Integration is carried out over the entire space. Taking account of Sommerfeld's radiation condition (see page 40):

$$\int (\vec{E}_1 \cdot \vec{J}_2 - \vec{E}_2 \cdot \vec{J}_1)\, dx = \int_S (\vec{H}_2\vec{E}_1 - \vec{H}_1\vec{E}_2) \cdot dS \underset{S \to \infty}{\rightarrow} 0 \tag{80}$$

which gives

$$\int_{R^3} \vec{E}_1 \cdot \vec{J}_2\, dx = \int_{R^3} \vec{E}_2 \cdot \vec{J}_1\, dx \tag{81}$$

This is the reciprocity theorem, which can be extended in various ways. It is also possible to establish:

$$\int \vec{E}_1 \cdot \vec{J}_2\, dx = -\int \vec{E}_1 \cdot \vec{J}_2\, dx \tag{82}$$

D: Maxwell's equation—mathematical aspects

1: Electromagnetic theory

As with other branches of theoretical physics, it is in the abstract formulation of Maxwell's theory that the basis for the most remarkable progress in electromagnetism is to be found. The potential contribution of the resolution of certain well defined mathematical problems to the description of e.m. phenomena had already been grasped by Hertz, whose axiomatic viewpoint (adopted in Stratton's excellent treatise [34]) is at the origin of this body of doctrine, which eventually became the mathematical theory of e.m. waves.

From this viewpoint, it is interesting to adopt as the starting point more general equations, justified simply by the practical consequences.

2: Magnetic distribution

Maxwell's equations can be made completely symmetrical by the introduction—quite gratuitously, for the present—of imaginary magnetic charges and current, with respective densities τ and $\vec{M}$ related by the continuity relationship

$$\operatorname{div} \vec{M} + i\omega\tau = 0 \tag{83}$$

and such that in a perfect medium:

$$\begin{aligned} \operatorname{curl} \vec{E} + i\omega\mu\vec{H} &= -\vec{M} \\ \operatorname{curl} \vec{H} - i\omega\varepsilon\vec{E} &= \vec{J} \\ \operatorname{div} \vec{E} &= \frac{1}{\varepsilon}\rho \\ \operatorname{div} \vec{H} &= \frac{1}{\mu}\tau \end{aligned} \tag{84}$$

A preliminary application is the complete duality, thus expressed, between electrical and magnetic properties. An appropriate exchange between electrical and magnetic magnitudes leaves Maxwell's equations invariant, in the form:

$$\begin{aligned} \vec{E} &\to \pm\eta\vec{H} \\ \vec{H} &\to \mp\frac{1}{\eta}\vec{E} \end{aligned} \tag{85}$$

where the constant

$$\eta = \sqrt{\frac{\mu}{\varepsilon}} \tag{86}$$

has the dimensions of an impedance, and is an extremely useful characteristic of the medium. *In vacuo*:

$$\eta_0 = \sqrt{\frac{\mu_0}{\varepsilon_0}} = 120\pi \simeq 377\Omega \tag{87}$$

So any solution to a propagation problem corresponds to the solution of a problem of this kind.

Examples

- *TE* and *TH* waves.
- Radiation from a Hertzian doublet and an elementary current loop.
- Radiation from a dipole and a half-wave slot.
- Diffraction by complementary screens (Babinet's principle).

3: Bicomplex field

(a) Definition

The e.m. correspondence suggests a compact representation of the e.m. field that can perform a number of services. The idea of using a complex combination, such as $E + jH$ (in Gauss's symmetrical units) is very old (Silberstein, 1907), but has not been very frequently used ([3], [9]). The reason for this is that the study of the sinusoidal case with fields that are themselves complex becomes ambiguous. The standard artifice of the double sign $\pm$ allows the difficulty to be removed. The use of commutative bicomplex algebra, however, with four units, l, i, j, ij, also solves the problem (see Appendix 1).

Taking into account the units adopted the vector

$$\vec{C} = \vec{E} + j\eta\vec{H} \quad (j = \sqrt{-1}) \tag{88}$$

is formed, based on the fields $\vec{E}$ and $\vec{H}$, which are themselves complex (written with $i = \sqrt{-1}$). For convenience, the field thus constructed is called bicomplex.

The densities of bicomplex charges and currents are defined in the same way:

$$\chi = \eta\rho + j\tau \tag{89}$$

$$\vec{K} = \eta\vec{J} + j\vec{M} \tag{90}$$

The extension of the strictly linear vector calculus is immediate. System (84) is replaced by:

$$\text{curl } \vec{C} - ijk\vec{C} = j\vec{K} \tag{91}$$

$$\text{div } \vec{C} = v\chi \tag{92}$$

with

$$\text{div } \vec{K} + i\omega\chi = 0 \tag{93}$$

Note that v is the velocity of propagation in the medium (e, μ):

$$v = \frac{1}{\sqrt{\varepsilon\mu}} = \frac{\omega}{k} \tag{94}$$

Note

- (92) is simply a consequence of (91) and (93).
- (93) means that only the bicomplex current $\vec{K}$ is involved in the reasoning.
- In the practical resolution of a problem, splitting up (88) allows the usual fields to be recovered. In the same way as an ordinary complex field has a real part and an imaginary part, so a bicomplex field has an electrical part and a magnetic part.

(b) Duality

(91) shows immediately that for any solution $\vec{C}$ to a problem there is a corresponding solution $j\vec{C}$ (or $-j\vec{C}$) linked to the sources $j\chi$ and $j\vec{K}$ (with $-j\chi$

and $-j\vec{K}$ respectively). So, returning to the usual quantities, ρ, τ, $\vec{J}$, $\vec{M}$, $\vec{E}$, $\vec{H}$, the following are associated with them respectively: $-\tau/\eta$, $\eta\rho$, $-\vec{M}/\eta$, $\eta\vec{J}$, $-\eta\vec{H}$, $\vec{E}/\eta$.

(c) Propagation equation

The curls of the two members of (91) lead to:

$$\begin{aligned}\text{curl curl } \vec{C} - ijk \text{ curl } \vec{C} &= \text{curl curl } \vec{C} - ijk(ijk\vec{C} + j\vec{K}) = j \text{ curl } \vec{K} \\ &= \text{curl curl } \vec{C} - k^2\vec{C} = -ik\vec{K} + j \text{ curl } \vec{K}\end{aligned} \tag{95}$$

But taking into account (92):

$$\text{curl curl } \vec{C} = \text{grad div } \vec{C} - \Delta\vec{C} = v \text{ grad } \chi - \Delta\vec{C} \tag{96}$$

Then:

$$\Delta\vec{C} + k^2\vec{C} = ik\vec{K} - j \text{ curl } \vec{K} + v \text{ grad } \chi \tag{97}$$

or else, according to (93):

$$\Delta\vec{C} + k^2\vec{C} = ik\vec{K} - j \text{ curl } \vec{K} + \frac{i}{k} \text{ grad div } \vec{K} \tag{98}$$

This leads to a single Helmholtz equation.

By separating the electrical and magnetic parts, the basic equations are recovered:

$$\Delta\vec{E} + k^2\vec{E} = \frac{1}{\varepsilon} \text{ grad } \rho + i\omega\mu\vec{J} + \text{curl } \vec{M} \tag{99}$$

$$= i\omega\mu\vec{J} + \frac{i}{\omega\varepsilon} \text{ grad div } \vec{J} + \text{curl } \vec{M} \tag{100}$$

$$\Delta\vec{H} + k^2\vec{H} = \frac{1}{\mu} \text{ grad } \tau + i\omega\varepsilon\vec{M} - \text{curl } \vec{J} \tag{101}$$

$$= i\omega\varepsilon\vec{M} + \frac{i}{\omega\mu} \text{ grad div } \vec{M} - \mathit{curl}\ \vec{J} \tag{102}$$

Note

In rectangular cartesian coordinates, ΔC is the vector with components

$$\Delta C_k = \Delta E_k + j\eta\Delta H_k \tag{103}$$

written with the usual Laplacian operator.

4: Use of distribution theory

(a) Summary of notation

Up to this stage, the validity of calculations has not been questioned. Sources and fields were, implicitly, assumed to be distributed with a regularity that ensured the continuity of the various derivatives. But in theory and application, situations in which the fields are discontinuous over surfaces are frequent, as in the case of finite bodies in a medium of different electrical characteristics. Maxwell's equations must be completed by boundary conditions. The theory of distributions provides great assistance, and although relatively old (1947), it has hardly been used in e.m. theory.

The notation used is given in Appendix 3. Let S be a regular surface—that is, at each point y it has an oriented unit normal $\vec{n}$, allowing a + face to be distinguished from a − face. Consider a scalar $a(x)$ (for example, a field component), which has two finite limits $a(y^+)$ and $a(y^-)$, depending on whether x tends towards y on the + side or the − side. The discontinuity across S will be expressed thus:

$$[a]_S = a(y^+) - a(y^-) \tag{104}$$

For a vector field $\vec{A}(x) = (A_1(x), A_2(x), A_3(x))$ the vector discontinuity will be denoted in the same way:

$$[A]_S = ([A_1]_S, [A_2]_S, [A_3]_S) \tag{105}$$

The simple layer distribution of density $f(y)$ over S is expressed $f\delta_S$. It is defined, functionally, by the complex number

$$\langle f\delta_S, \varphi\rangle = \int_S f(y)\varphi(y)\, dS \tag{106}$$

associated with each of the functions $\varphi(x)$ of a suitable vector space $\mathscr{D}$ (Appendix 3).

The double layer distribution of density $g(y)$ is denoted $\partial_n(g\delta_S)$ and is such that:

$$\langle \partial_n(g\delta_S), \varphi\rangle = -\int_S g(y)\, \partial_n\varphi(y)\, dS \tag{107}$$

The distribution associated with the locally integrable scalar $f(x)$ is denoted **f**. The vector distribution **A** associated with the field $A(x)$ will be considered:

$$\vec{\mathbf{A}} = (\mathbf{A}_1, \mathbf{A}_2, \mathbf{A}_3) \tag{108}$$

The following relationships (Appendix 3) are found:

$$\operatorname{div} \vec{\mathbf{A}} = \operatorname{div} \vec{A} + \vec{n}\cdot[\vec{A}]_S\delta_S \tag{109}$$

$$\operatorname{curl} \vec{\mathbf{A}} = \operatorname{curl} \vec{A} + \vec{n} \wedge [\vec{A}]_S\delta_S \tag{110}$$

and are valid not only in the complex representation but also in the bicomplex.

(b) Application to the e.m. field

Consider Maxwell's equations as relationships between the distributions associated with various electromagnetic magnitudes:

$$\begin{aligned} \text{curl}\ \vec{\mathbf{E}} + i\omega\vec{\mathbf{B}} &= -\vec{\mathbf{M}} \\ \text{curl}\ \vec{\mathbf{H}} - i\omega\vec{\mathbf{D}} &= \vec{\mathbf{J}} \\ \text{div}\ \vec{\mathbf{D}} &= \rho \\ \text{div}\ \vec{\mathbf{B}} &= \tau \end{aligned} \tag{111}$$

with

$$\begin{aligned} \text{div}\ \vec{\mathbf{J}} + i\omega\rho &= 0 \\ \text{div}\ \vec{\mathbf{M}} + i\omega\tau &= 0 \end{aligned} \tag{112}$$

Any surface of field discontinuity reveals simple scalar or vector layers:

$$\begin{aligned} \vec{\mathbf{J}} &= \vec{J} + \vec{J}_S\delta_S \\ \vec{\mathbf{M}} &= \vec{M} + \vec{M}_S\delta_S \\ \boldsymbol{\rho} &= \rho + \rho_S\delta_S \\ \boldsymbol{\tau} &= \tau + \tau_S\delta_S \end{aligned} \tag{113}$$

where, according to (109) and (110):

$$\begin{aligned} \vec{J}_S &= n \wedge [\vec{H}]_S \\ \vec{M}_S &= -\vec{n} \wedge [\vec{E}]_S \\ \rho_S &= \vec{n}\cdot[\vec{D}]_S \\ \tau_S &= \vec{n}\cdot[\vec{B}]_S \end{aligned} \tag{114}$$

In the same way for a discontinuity surface for currents:

$$\begin{aligned} \rho_S &= -\frac{1}{i\omega}\vec{n}\cdot[\vec{J}]_S \\ \tau_S &= -\frac{1}{i\omega}\vec{n}\cdot[\vec{M}]_S \end{aligned} \tag{115}$$

(c) Usual boundary conditions

In physical reality there is no magnetic repartition or any repartition of surface electric current, only a repartition of the surface charges.

The equations (114) can, therefore, be reduced to:

$$\begin{aligned} \vec{n} \wedge [\vec{E}]_S &= 0 \\ \vec{n} \wedge [\vec{H}]_S &= 0 \\ \vec{n} \wedge [\vec{B}]_S &= 0 \end{aligned} \tag{116}$$

The continuity of the tangential components of the electric and magnetic fields and of the normal component of magnetic induction are re-established.

The model of the perfect conductor is fundamental. S limits a perfect conductor in a perfect dielectric (ε, μ), towards which the normal is oriented. The conductor is empty of fields, charges and currents, a limiting case which implies the existence of a surface covering of current, $\vec{J}_S$.

The following equations can be written:

$$\begin{aligned} \vec{E}(y^+) &= \vec{E}_S \quad \text{with} \quad \vec{E}(y^-) = 0 \\ \vec{H}(y^+) &= \vec{H}_S \quad \text{with} \quad \vec{H}(y^-) = 0 \end{aligned} \tag{117}$$

It follows that:

$$\begin{aligned} \vec{n} \wedge \vec{E}_S &= 0 \\ \vec{n} \wedge \vec{H}_S &= \vec{J}_S \\ \vec{n} \cdot \vec{E}_S &= \frac{1}{\varepsilon} \rho_S \\ \vec{n} \cdot \vec{H}_S &= 0 \end{aligned} \tag{118}$$

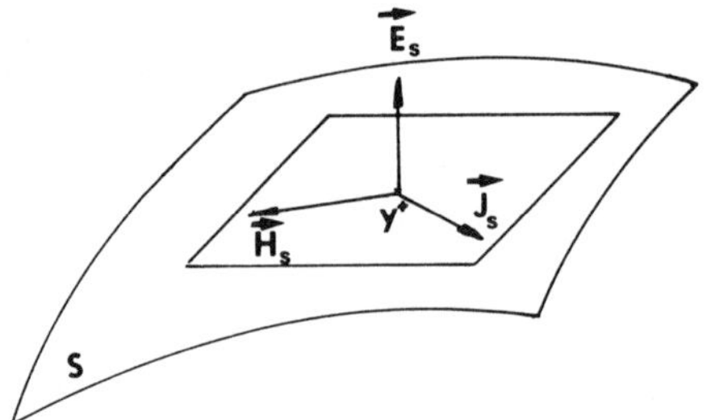

Fig. 2. *Boundary conditions in a perfect conductor.*

The above boundary conditions are summarized as follows:

- On the surface of a perfect conductor, the external electric field is normal and the external magnetic field is tangential.
- The electric field measures the density of charge.
- The magnetic field measures the density of current (rotates $\vec{H}_S$ by $+\pi/2$ about $\vec{n}$).

At each instant the vectors $\mathscr{E}_S$, $\mathscr{H}_S$, $\mathscr{J}_S$ (in alphabetical order) are orthogonal and related by the right-hand screw rule.

Non-regular surfaces: Meixner's condition

The above does not apply in the frequent case where S presents singularities like edges and wedges, for example. Not only can the fields become infinite, but it will also be seen that there is a risk of obtaining multiple solutions. Meixner's uniqueness condition states that in any neighbourhood around a singularity, the average localized energy (63) is finite, setting a limit on the rapidity of increase of the field components.

Examples

• Ridge: Consider a cylinder with, as its axis, an element dz of the ridge. In semipolar coordinates, the element of volume is $dx = \rho\, d\rho\, d\varphi\, dz$.

For any field component in the form $f(\rho)g(\varphi, z)$, the following can be expressed:

$$\int_0^a |f(\rho)|^2 \rho\, d\rho < \infty \tag{119}$$

a condition obeyed for:

$$|f(\rho)| = \mathrm{O}(\rho^{-\alpha}), \quad \alpha < 1 \tag{120}$$

• Peak: A sphere of radius a about the vertex is considered. In polar coordinates the volume element is $dx = r^2 \sin\theta\, dr\, d\theta\, d\varphi$ with a component $f(r)g(\theta, \varphi)$. The condition

$$\int_0^a |f(r)|^2 r^2\, dr < \infty \tag{121}$$

is obeyed for

$$|f(r)| = \mathrm{O}(r^{-\beta}), \quad \beta < \frac{3}{2} \tag{122}$$

(d) Sobolev's technique

This consists of introducing boundary or initial conditions into a problem of differential or partial differential equations, using an artificial discontinuity obtained by 'sweeping out' an appropriate domain—that is, by removing the values involved from it.

In electromagnetism, this provides the Huygens-Kirchoff representation of the field (equivalence theorems). Consider, as an example, limited sources creating a field $(\vec{E}, \vec{H})$ in an unlimited perfect dielectric (ε, μ). A regular closed surface S, surrounding the sources, separates a limited internal domain Ω' from the unlimited external domain Ω, towards which the normal is oriented. $(\vec{E}_S, \vec{H}_S)$ is the field on S.

In $\mathbf{R}^3$, the e.m. field is made up of the primary field at any point in Ω, and there is an identically zero field in Ω'. This gives rise to the discontinuities through the surface, with which, using (114), the superficial or secondary sources are associated:

$$\begin{aligned} \vec{J}_S &= n \wedge \vec{H}_S \\ \vec{M}_S &= -\vec{n} \wedge \vec{E}_S \\ \rho_S &= \varepsilon \vec{n} \cdot \vec{E}_S \\ \tau_S &= \mu \vec{n} \cdot \vec{H}_S \end{aligned} \tag{123}$$

The fact that the imaginary field can be entirely determined by secondary sources constitutes Huygens' principle (which will, in fact, be a theorem).

Exercises

• Find the results directly from (91) and (92), where the bicomplex sources and field are distributions. Verify that on every discontinuity surface of the field:

$$\vec{K}_S = -j\vec{n} \wedge [\vec{C}]_S$$
$$\chi_S = \frac{1}{v}\vec{n}\cdot[\vec{C}]_S \tag{124}$$

• As a calculation variant, the sweeping-out $1_\Omega\vec{C}$, $1_\Omega\vec{K}$, $1_\Omega\chi$ where 1_Ω is the characteristic function of the domain Ω, will be carried out:

$$1_\Omega(x) = 1, \quad x \in \Omega$$
$$= 0, \quad x \in \Omega' \tag{125}$$

(For the present, 1_Ω is not defined on S.)

The following relationship will be used:

$$\text{grad } 1_\Omega = -\vec{n}\delta_S \tag{126}$$

• Verify that, for the distributions, the following conservation relationships apply:

$$\text{div } (\vec{J}_S\delta_S) + i\omega(\rho_S\delta_S) = 0$$
$$\text{div } (\vec{M}_S\delta_S) + i\omega(\tau_S\delta_S) = 0 \tag{127}$$

Note

The above justifies the introduction of magnetic layers. These are necessary here to account for the discontinuities imposed on the field. If S were a surface material, for example, of the sort that limits a perfect conductor, the boundary conditions (see (118)) would be those encountered where there are no such layers.

E: Radiation of electromagnetic sources

1: Source and radiation field

The major problem, and the one that encompasses all others, with transmitting antennas, is that of expressing the radiation field using the sources, charges and associated currents which it supports. This is restricted, and the term 'radiation field' implies the solution of an external problem. Maxwell's equations, which express the interaction between fields, charges and currents, including possible constraints (boundary conditions, for example), solve the problem. The hypothesis that currents are known in advance—that is, independently of the field created (with sources understood in the strictest sense of the word)—gives rise in a perfect medium to an immediate solution. It so happens that in most antenna problems, where only the far field is of interest, this hypothesis provides an excellent approximation. Otherwise the currents (with sources considered in the broad sense) can be considered as a calculation intermediary. This is the case with diffraction problems.

Finally, the equations (97) or (98) can be used as the starting point:

$$\Delta\vec{C} + k^2\vec{C} = \vec{S} \tag{128}$$

where the second term,

$$\begin{aligned}\vec{S} &= ik\vec{K} - j\,\text{curl}\,\vec{K} + v\,\text{grad}\,\chi\\ &= ik\vec{K} - j\,\text{curl}\,\vec{K} + \frac{i}{k}\,\text{grad div}\,\vec{K} = \left[\frac{i}{k}(\text{grad div} + k^2) - j\,\text{curl}\right]\vec{K}\end{aligned} \tag{129}$$

will represent the composite sources, both electric and magnetic.

Each cartesian component in the desired field $\vec{C}$ therefore satisfies the scalar Helmholtz equation,

$$\Delta U + k^2 U = f \tag{130}$$

a discussion of which can be found in references [31], [36], [26].

2: Scalar Helmholtz equation

(a) Use of distributions

There are many advantages in this. The source distribution **f** directly represents all practical situations: sources with volume density, simple or multiple layers, curvilinear antennas, multipolar sources, etc. Equation (130) is treated as a particular convolution equation. Finally, it does not raise any problems of discontinuity and differentiation. Therefore, if S is a regular closed surface, the exterior of which is Ω, the sweeping-out of the interior simply gives

$$\Delta(1_\Omega U) = 1_\Omega \Delta U - \partial_n U \cdot \delta_S - \partial_n(U\delta_S) \tag{131}$$

—that is, the combination of the usual swept-out Laplacian, a simple layer of density $-\partial_n U$ and a double layer of density U. The boundary conditions, such that:

$$U_S = g, \quad \partial_n U_s = h \tag{132}$$

are reduced to a simple modification of the second term (Sobolev's method):

$$\Delta U + k^2 U = f + h\delta_S + \partial_n(g\delta_S) \quad (x \in \Omega) \tag{133}$$

(b) Homogeneous equation

This is the specific case where:

$$\Delta U + k^2 U = 0 \tag{134}$$

Its importance is based on:

Theorem 1: The difference between any two solutions of the complete equation (130) is a solution to the homogeneous equation.

Therefore the general solution to a problem is the sum of a particular solution to the equation and the general solution of the homogeneous equation.

The homogeneous equation, however, always admits the trivial solution $U = 0$. If it admits a non-trivial solution $U \neq 0$, then it admits an infinite number, such as cU (c = constant), any partial derivative, any translation, etc.

The fundamental result of this is:

Theorem 2: The necessary and sufficient condition for the complete equation to admit only a single solution is that the homogeneous equation must admit only the trivial solution.

(c) Elementary solutions

This is the name given to any solution of the particular equation:

$$\Delta G + k^2 G = \delta \tag{135}$$

It is known that in $\mathbf{R}^3$ (see Appendix 3)

$$G(x) = -\frac{e^{-ik|x|}}{4\pi|x|} = -\frac{e^{-ikr}}{4\pi r}, \quad (r = |x|) \tag{136}$$

and its conjugate:

$$G^*(x) = -\frac{e^{ikr}}{4\pi r} \tag{137}$$

are elementary solutions.

Their physical meaning is simple. According to the rule given in (24),

$$\mathscr{R}\mathrm{e}\,\{G(x)\,e^{i\omega t}\} = -\frac{\cos(\omega t - kr)}{4\pi r} \tag{138}$$

which represents a divergent spherical wave (propagating towards infinity). With (137) there would be a convergent spherical wave (towards the origin).

The identities

$$\begin{aligned} \frac{dG}{dr} + ikG &= -\frac{G}{r} \\ \frac{dG^*}{dr} - ikG^* &= -\frac{G^*}{r} \end{aligned} \tag{139}$$

allow them to be distinguished by their asymptotic behaviour. It can therefore be observed that:

$$\frac{dG}{dr} + ikG = \mathrm{O}\left(\frac{1}{r}\right)$$
$$\frac{dG^*}{dr} - ikG^* = \mathrm{O}\left(\frac{1}{r}\right) \tag{140}$$

the generalization of which will lead to an important uniqueness criterion (Sommerfeld's radiation condition).

Note

• In $\mathbf{R}^2$ the elementary solution would be:

$$G(x) = \frac{i}{4} H_0^{(2)}(kr) \tag{141}$$

and generally, in $\mathbf{R}^n$ (see [31]):

$$G(x) = \frac{i}{4}\left(\frac{k}{2\pi r}\right)^{\pi/2-1} H_{n/2-1}^{(2)}(kr) \tag{142}$$

($r = |x|$; H: Hankel function.)

• It is clear that for any U, solution to the homogeneous equation, $G + U$ continues to satisfy (135). If U is calculated in such a way as to ensure a boundary condition such that $G + U$, or its normal derivative, is zero on a given surface, Green's function applies. The elementary solution G is often, mistakenly, given this name.

(d) Solution of the complete equation $\Delta \mathbf{U} + k^2\mathbf{U} = \mathbf{f}$

The importance of an elementary solution is explained by:

Theorem 3: The convolution of an elementary solution with the second member of the complete equation is the particular solution of the latter.

This solution:

$$U = G * \mathbf{f} \tag{143}$$

reduces to a usual function for any distribution $\mathbf{f}$ of a limited extent (see Appendix 3).

If, in particular, f is an ordinary density, then:

$$U(x) = \int_{\mathbf{R}^3} G(x - x')f(x')\,dx' = -\frac{1}{4\pi}\int f(x')\frac{e^{-ikr}}{r}\,dx' \tag{144}$$

where

$$r = |x - x'|$$

This can be recognized as the expression for a retarded potential.

The problem of uniqueness will be considered later.

3: Calculating the radiation field

(a) Basic formula for radiation from given sources

By bicomplex vectorial extension of (143), the convolution

$$\vec{C} = G * \vec{\mathbf{S}} = \vec{\mathbf{S}} * G \tag{145}$$

is a solution to Helmholtz's equation (127).

Taking into account expression (129) of $\vec{\mathbf{S}}$, this becomes explicitly:

$$\vec{C} = \left\{\left[\frac{i}{k}(\text{grad div} + k^2) - j\,\text{curl}\right]\vec{\mathbf{K}}\right\} * G \tag{146}$$

$$= \left(\frac{i}{k}\,\text{grad div}\,\vec{\mathbf{K}} + ik\,\vec{\mathbf{K}} - j\,\text{curl}\,\vec{\mathbf{K}}\right) * G \tag{147}$$

This basic result, which allows the field created by the most general source distribution to be calculated, is the basis for all radiation calculations.

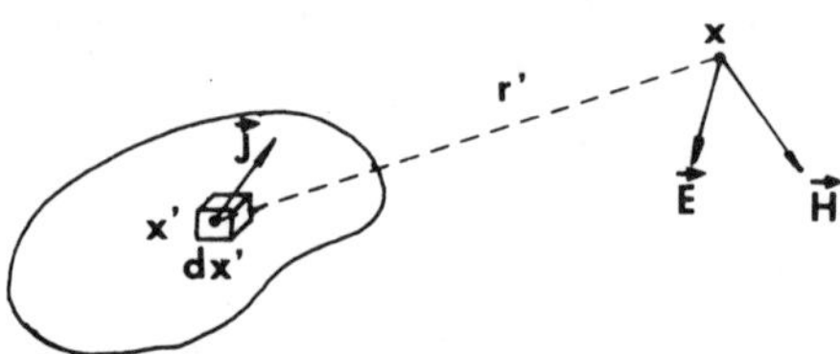

Fig. 3 *Radiation from electrical sources.*

(b) Various equivalent expressions

From the following property of the derivative of convolutions

$$\partial_i(\mathbf{K}_j * G) = \partial_i\mathbf{K}_j * G = \mathbf{K}_j * \partial_i G \tag{148}$$

this expression can be derived:

$$\vec{C} = \left[\frac{i}{k}(\text{grad div} + \text{k}^2) - j\,\text{curl}\right](\vec{\mathbf{K}} * G) \tag{149}$$

which suggests the possible use, as a calculational intermediary, of a bicomplex potential (see below).

It may be helpful, in some calculations, to use the charge χ from the expression (128) of $\vec{\mathbf{S}}$:

$$\vec{C} = v\,\text{grad}\,\chi * G + (ik - j\,\text{curl})\,\vec{\mathbf{K}} * G \tag{150}$$

Note

- The distributions can be used for derivatives of any order, automatically taking into account the discontinuities.
- With the notation given in Appendix 3 (page 212) the following expressions

can also be used for radiation fields:

$$\vec{C} = \frac{i}{k}\,\text{grad}\,(\vec{\mathbf{K}} * \text{grad}\,G) + ik\vec{\mathbf{K}} * G + j\vec{\mathbf{K}} \overset{*}{\wedge} \text{grad}\,G \tag{151}$$

$$= v\chi * \text{grad}\,G + ik\vec{\mathbf{K}} * G + j\vec{\mathbf{K}} \overset{*}{\wedge} \text{grad}\,G \tag{152}$$

• In practical calculations it is important to remember that in $G(x - x')$ the derivatives relate to the components of vector x.

(c) Potentials

Expressions (149) and (150) lead naturally to the appearance of bicomplex scalar and vector potentials:

$$V = V_e + j\eta V_m = -v\chi * G \tag{153}$$

$$\vec{A} = \vec{A}_e + j\eta\vec{A}_m = -\frac{1}{v}\vec{K} * G \tag{154}$$

The sole object of the coefficients is to provide a mode of expression that is consistent with those for the usual electric and magnetic potentials:

$$V_e, \vec{A}_e \quad \text{and} \quad V_m, \vec{A}_m$$

Lorentz's equation can be verified immediately:

$$\text{div}\,\vec{A} + i\omega\varepsilon\mu V = 0 \tag{155}$$

from which Helmholtz's potential equations are found:

$$\Delta V + k^2 V = -v\chi \tag{156}$$

$$\Delta\vec{A} + k^2\vec{A} = -\frac{1}{v}\vec{K} \tag{157}$$

The field can be expressed as:

$$\vec{C} = -\text{grad}\,V - i\omega\vec{A} + jv\,\text{curl}\,\vec{A} \tag{158}$$

or as

$$\vec{C} = \frac{1}{i\omega\varepsilon\mu}(\text{grad div}\,\vec{A} + k^2\vec{A}) + jv\,\text{curl}\,\vec{A} \tag{159}$$

In the same way that (149) expresses the field $\vec{C}$ on the basis of the single current $\vec{K}$, (159) expresses it as a function of the single vector potential $\vec{A}$.

It should be noted that the calculation with distributions attaches far less importance to the potentials than the traditional formulation.

(d) The conventional field expressions

The breakdowns into electrical and magnetic parts in (88), (89) and (90) give the following more familiar results:

$$\vec{E} = \frac{i}{\omega\varepsilon}(\text{grad div} + k^2)(\vec{J} * G) + \text{curl}\,(\vec{M} * G)$$
$$\vec{H} = \frac{i}{\omega\mu}(\text{grad div} + k^2)(\vec{M} * G) - \text{curl}\,(\vec{J} * G) \tag{160}$$

as a function of the single electric and magnetic currents. In some problems, the corresponding charges can also be used:

$$\vec{E} = \frac{1}{\varepsilon}\,\text{grad}\,(\rho * G) + i\omega\mu(\vec{J} * G) + \text{curl}\,(\vec{M} * G)$$
$$\vec{H} = \frac{1}{\mu}\,\text{grad}\,(\tau * G) + i\omega\varepsilon(\vec{M} * G) - \text{curl}\,(\vec{J} * G) \tag{161}$$

The following is a consequence of the preceding:

$$\vec{E} = \frac{1}{\omega\varepsilon}\,\text{grad}\,(\vec{J} \overset{*}{\cdot} \text{grad}\, G) + i\omega\mu(\vec{J} * G) - \vec{M} \overset{*}{\wedge} \text{grad}\, G$$
$$\vec{H} = \frac{1}{\omega\mu}\,\text{grad}\,(\vec{M} \overset{*}{\cdot} \text{grad}\, G) + i\omega\varepsilon(\vec{M} * G) + \vec{J} \overset{*}{\wedge} \text{grad}\, G \tag{162}$$

and the first terms of the second terms can be replaced respectively, by $\frac{1}{\omega}\rho * \text{grad}\, G$ and $\frac{1}{\mu}\tau * \text{grad}\, G$.

Finally, with the usual potentials:

$$V_e = -\frac{1}{\varepsilon}\rho * G, \quad \vec{A}_e = -\mu\vec{J} * G$$
$$V_m = -\frac{1}{\mu}\tau * G, \quad \vec{A}_m = -\varepsilon\vec{M} * G \tag{163}$$

$$\vec{E} = -\text{grad}\, V_e - i\omega\vec{A}_e - \frac{1}{\varepsilon}\,\text{curl}\,\vec{A}_m$$
$$= \frac{1}{i\omega\varepsilon\mu}(\text{grad div}\,\vec{A}_e + k^2\vec{A}_e) - \frac{1}{\varepsilon}\,\text{curl}\,\vec{A}_m \tag{164}$$

$$\vec{H} = -\text{grad}\, V_m - i\omega\vec{A}_m + \frac{1}{\mu}\,\text{curl}\,\vec{A}_e$$
$$= \frac{1}{i\omega\varepsilon\mu}(\text{grad div}\,\vec{A}_m + k^2\vec{A}_m) + \frac{1}{\mu}\,\text{curl}\,\vec{A}_e \tag{165}$$

Note

The formulae in this paragraph are almost all standard, and expressed here with ordinary densities of current and charge. The above justifies their extension to source distributions.

4: Application to antennas

(a) Simple antennas

The most important illustration of the above formulae is their application to sources with density. A simple antenna is the limited extent of a distribution of electric currents with a density initially considered as a volume. The formulae in (160) give, for any point x external to the antenna:

$$\vec{E}(x) = \frac{i}{\omega\varepsilon}(\text{grad div} + k^2)\int \vec{J}(x')G(x - x')\,dx' = \frac{1}{4\pi i\omega\varepsilon}(\text{grad div} + k^2)\int \vec{J}(x')\frac{e^{-ikr'}}{r'}\,dx' \quad (166)$$

$$\vec{H}(x) = \frac{1}{4\pi}\,\text{curl}\int \vec{J}(x')\frac{e^{-ikr'}}{r'}\,dx' \quad (167)$$

$$(r' = |x - x'|)$$

With (162), the following is found even more directly:

$$\vec{E}(x) = \frac{1}{4\pi i\omega\varepsilon}\int\left\{\text{grad}\left[\vec{J}(x')\cdot\text{grad}\frac{e^{-ikr'}}{r'}\right] + k^2\vec{J}(x')\frac{e^{-ikr'}}{r'}\right\}dx' \quad (168)$$

$$\vec{H}(x) = -\frac{1}{4\pi}\int \vec{J}(x') \wedge \text{grad}\frac{e^{-ikr'}}{r'}\,dx' \quad (169)$$

Most simple antennas are conductors that can be assumed perfect. An antenna of this sort is therefore the support of a surface density of electric current. The expressions for radiation fields are the same, except that they must be 'read' with the integrals extended to the surface of the antenna, and with the differential element $J_s\,dS$ playing the part of $J\,dx'$. The same formulae can also be applied to the extremely common case of the wire antenna. This is treated as a curve along which an electric current of intensity $I(x')$ is spread.

The integrals (166) to (169) are then curvilinear and $I(x')\,\vec{ds}$ replaces $\vec{J}(x')\,dx$.

(b) Dipolar sources

(1) Origin and role Hertz's sphere discharger (the first transmitting antenna!) is generally treated as an oscillating electrical dipole, in which the enormous capacity of the spheres has the effect of keeping the distribution of current intensity uniform along the radiating rectilinear element. This is why any rectilinear wire antenna with this current characteristic (an ordinary alternating current without propagation) is called a Hertzian doublet. This name avoids the ambiguity of the term dipole, which is used differently in the study of antennas.

The mathematical model is the source distribution localized, for example, at the origin:

$$\vec{\mathbf{J}} = \vec{J}\delta \tag{170}$$

(32) associated with this the distribution charge:

$$\rho = -\frac{1}{i\omega}\operatorname{div}\vec{\mathbf{J}} = -\frac{1}{i\omega}\vec{J}\cdot\operatorname{grad}\delta \tag{171}$$

revealing the derivative of δ in the direction of the current, that is, effectively a dipole in that direction.

A source of this type gives rise to the simplest calculations. It has varied applications, which will be discussed in Part 2, and although its theoretical importance, which was considerable in the pioneering period, has greatly diminished, this is mainly because of the development of technique towards ever shorter waves.

The theory of distributions can provide it with an interesting role. This relates to the important theorem [13]:

Any distribution with limited extent is the limit of a series of finite linear combinations of Dirac delta functions.

$$\mathbf{S} = \lim_{n\to\infty} \mathbf{S}_n$$

$$\mathbf{S}_n = \sum_i \lambda_{n,i}\delta_{(x_{n,i})}$$

This, specifically, is how multipoles are defined. The vectorial transposition is evident. The continuity of the convolution

$$\vec{C} = G * \vec{\mathbf{S}} = \lim_{n\to\infty} G * \vec{\mathbf{S}}_n \tag{172}$$

means that the above comment and the following calculation are of particular interest.

(2) Radiation from a reduced dipolar source Consider the source:

$$\vec{\mathbf{s}} = \vec{s}\delta \tag{173}$$

where $\vec{s}$ is a unit vector in O. (151) can simply be used directly.

There is no problem with the calculations. A number of specific points are given below. With the unit vector of $\vec{x}$:

$$\vec{u} = \frac{\vec{x}}{r}, \quad r = |\vec{x}| \tag{174}$$

and the ordinary differentiation of G with respect to r. This gives:

$$\begin{aligned}\vec{\mathbf{s}} * G &= G\vec{s}\\ \vec{\mathbf{s}} \overset{*}{\cdot} \operatorname{grad} G &= G'\vec{s}\cdot\vec{u}\\ \operatorname{grad}(\vec{\mathbf{s}} \overset{*}{\cdot} \operatorname{grad} G) &= r\left(\frac{G'}{r}\right)'(\vec{s}\cdot\vec{u})\vec{u} + \frac{G'}{r}\vec{s}\\ \vec{\mathbf{s}} \overset{*}{\wedge} \operatorname{grad} G &= G'\vec{s}\wedge\vec{u}\end{aligned} \tag{175}$$

It is then easy to obtain the field:

$$\vec{C}_0(x;\vec{s}) = ik\left[\left(-1 + \frac{3i}{kr} + \frac{3}{k^2r^2}\right)(\vec{s}\wedge\vec{u})\wedge\vec{u} + \left(\frac{2i}{kr} + \frac{2}{k^2r^2}\right)\vec{s} - j\left(1 - \frac{i}{kr}\right)\vec{s}\wedge\vec{u}\right]G \tag{176}$$

which reveals the asymptotic behaviour ($r \to \infty$):

$$\vec{C}_0(x;\vec{s}) = -ik[(\vec{s}\wedge\vec{u} + j\vec{s})\wedge\vec{u}]G + o\left(\frac{1}{r}\right) \tag{177}$$

The symbol $o\left(\frac{1}{r}\right)$ is understood vectorially here.

(3) **Any dipolar source** Consider the dipolar source of current $\vec{K}$ (bicomplex) located at a point x':

$$\vec{\mathbf{K}} = \vec{K}\delta_{(x')} \tag{178}$$

By 'homothesis' and translation x', the following expression can be used: $\vec{C}_0(x - x'; \vec{K})$, where in the second member of (176) $\vec{u}$ is replaced by the unit vector:

$$\vec{u}' = \frac{\vec{x} - \vec{x}'}{r'}, \quad r' = |\vec{x} - \vec{x}'| \tag{179}$$

Example
Consider the particular case of the 'vertical' dipole ($\vec{s} = \vec{e}_3$) at the origin:

$$\vec{\mathbf{K}} = K\delta\vec{e}_3 = (\eta J + jM)\delta\vec{e}_3 \tag{180}$$

The calculation in polar coordinates (r, θ, φ) is immediate:

$$\vec{s} = \vec{e}_3 = (\cos\theta, -\sin\theta, 0), \quad \vec{u} = \vec{e}_r = (1, 0, 0) \tag{181}$$

It follows, according to (176) and after separation of the fields $\vec{E}$ and $\vec{H}$:

$$E_r = i\omega\mu J\left[\left(\frac{2i}{kr} + \frac{2}{k^2r^2}\right)\cos\theta\right]G, \qquad H_r = i\omega\varepsilon M\left[\left(\frac{2i}{kr} + \frac{2}{k^2r^2}\right)\cos\theta\right]G$$

$$E_\theta = i\omega\mu J\left[\left(-1 + \frac{i}{kr} + \frac{1}{k^2r^2}\right)\sin\theta\right]G, \tag{182}$$

$$H_\theta = i\omega\varepsilon M\left[\left(-1 + \frac{i}{kr} + \frac{1}{k^2r^2}\right)\sin\theta\right]G$$

$$E_\varphi = ikM\left[\left(1 - \frac{i}{kr}\right)\sin\theta\right]G, \qquad H_\varphi = -ikJ\left[\left(1 - \frac{i}{kr}\right)\sin\theta\right]G$$

These formulae can be used for the detailed consideration of the fields radiated by the vertical Hertzian doublet and the correlative magnetic doublet respectively (the latter is a small current loop with axis Ox_3).

(c) Multipolar sources

Consider, at the origin, a unit vector $\vec{\alpha}$. The following linear combination (of neighbouring and opposed elements of current) can be formed from (173):

$$\frac{1}{h}\vec{s}\,\delta_{(\vec{\alpha}h)} - \frac{1}{h}\vec{s}\,\delta$$

and converges, for $h \to 0$, towards the distribution current:

$$\vec{\mathbf{S}} = \vec{s}\,\partial_\alpha\delta \tag{183}$$

with which the distribution charge is associated:

$$\rho = -\frac{1}{i\omega}\operatorname{div}\vec{\mathbf{s}} = -\frac{1}{i\omega}\vec{s}\cdot\operatorname{grad}\partial_\alpha\delta = -\frac{1}{i\omega}\partial^2_{\alpha s}\delta \tag{184}$$

that is, a quadrupole.

Generally, this results in n unit vectors: $\vec{\alpha}_1, \ldots, \vec{\alpha}_n$, not necessarily all distinct, and the differentiation symbol with respect to their directions:

$$D^{(n)} = \partial^{(n)}_{\alpha_1,\ldots,\alpha_n} \tag{185}$$

The generalized current can be formed:

$$\vec{\mathbf{S}} = \vec{s}D^{(n)}\delta \tag{186}$$

to which corresponds, as above, a multipole of order 2^{n+1} (that is, a '2^{n+1}–pole').

The radiation field is found by applying rule (172) to (176):

$$\vec{C} = D^{(n)}\vec{C}_0 \tag{187}$$

Note

The compact form of expression used must not obscure the tensor character of multipoles. Each differentiation introduces a degree of variance.

Exercise
Apply Taylor's formula to $G(x - x')$. Deduce from this a representation of the radiation based on that of the multipoles at the origin.

(d) Huygens sources
Consider, once again, the schema described on page 23: a closed, regular surface S contains in its interior Ω' all the sources which will be referred to as primary. The object is to express the field $(\vec{E}, \vec{H})$ for any point in the domain Ω, external to S, based on the values $\vec{E}_S$, $\vec{H}_S$ on S.

Sobolev's technique (page 23) is equivalent to introducing into S the surface sources (124) known as secondary, or Huygens, sources. This gives rise immediately to:

$$1_\Omega \vec{C} = (\vec{n} \cdot C_S) * \operatorname{grad} G - ijk(\vec{n} \wedge \vec{C}_S) * G + (\vec{n} \wedge \vec{C}_S) \overset{*}{\wedge} \operatorname{grad} G \quad (188)$$

or, explicitly:

$$1_\Omega \vec{C} = \int_S [(\vec{n} \cdot \vec{C}_S) \operatorname{grad} G - ijk(\vec{n} \wedge \vec{C}_S) G + (\vec{n} \wedge \vec{C}_S) \wedge \operatorname{grad} G] \, dS \quad (189)$$

The 1_Ω factor ensures that the field $\vec{C}$ is found in Ω and a zero field is found in Ω'.

By separation, the classic formulae of Stratton and Chu are recovered:

$$1_\Omega \vec{E} = \int_S [(\vec{n} \cdot \vec{E}_S) \operatorname{grad} G + i\omega\mu(\vec{n} \wedge \vec{H}_S) G + (\vec{n} \wedge \vec{E}_S) \wedge \operatorname{grad} G] \, dS$$
$$(190)$$
$$1_\Omega \vec{H} = \int_S [(\vec{n} \cdot \vec{H}_S) \operatorname{grad} G - i\omega\varepsilon(\vec{n} \wedge \vec{E}_S) G + (\vec{n} \wedge \vec{H}_S) \wedge \operatorname{grad} G] \, dS$$

Note

- The same expressions apply for the internal field $1_{\Omega'} E$, $1_{\Omega'} H$, as long as the normal is oriented towards the interior at S.
- $G(x, x')$ is a function of the pair of points x, x'. The differentiations under the summation sign imply that the variable is x.

It is also possible, in (189), to use the derivative at x' simply by noting the change of sign:

$$\operatorname{grad} G(x - x') = \operatorname{grad}_x G = -\operatorname{grad}_x G \quad (191)$$

which will be written as $-\operatorname{grad}' G$.

In the far field ($kr' \gg 1$):

$$\operatorname{grad}' G = -\left(ik + \frac{1}{r'}\right) G\vec{u}' \simeq -ikG\vec{u}' \quad (192)$$

- It is easy to generalize (189) to a situation in which part of the sources would be situated in Ω, by adding the unswept-out contributions of $1_\Omega \vec{J}$ and $1_\Omega \vec{M}$.
- (189) can be considered as a formulation of Huygens' 'principle' (which is, in fact, a theorem), in the way it is understood by optical scientists, but which is

rather different from the way in which mathematicians see it. The formulae (190) are very important for the calculation of radiation through apertures (an approximation of optical physics).

(e) Plane sources (Figure 25)

The particular case of surface sources $\vec{J}_S$ and $\vec{M}_S$ in a plane S lends itself to simple considerations of symmetry which will be mentioned throughout this work. S separates the half-spaces Ω^- and Ω^+ and is oriented by the normal $\vec{n}$, directed towards Ω^+.

The point that is symmetric to a point x with respect to S will be called $\underline{x}$, and:

$$\underline{\vec{E}} = \vec{E}_t - \vec{E}_n \tag{193}$$

will be the symmetrical counterpart of any vector broken down into components parallel and orthogonal to S.

In addition, in equations (160) the field $\vec{E}_J(x)$, $\vec{H}_J(x)$ created at x by the single electric current will be distinguished from the field $\vec{E}_M(x)$, $\vec{H}_M(x)$ created by the single magnetic current.

The tensor properties in $\mathbf{R}^3$ of the grad div and curl operators (cf the old distinction between polar and axial vectors) lead to the symmetry properties:

$$\vec{E}_J(\underline{x}) = \underline{\vec{E}_J}(x), \qquad \vec{H}_J(\underline{x}) = -\underline{\vec{H}_J}(x) \tag{194}$$

$$\vec{E}_M(\underline{x}) = -\underline{\vec{E}_M}(x), \quad \vec{H}_M(\underline{x}) = \underline{\vec{H}_M}(x) \tag{195}$$

These are apparent, at least in the far field, in the asymptotic expressions (210) and (212).

On S itself, there is, in particular:

$$\vec{E}_{J.t} = \vec{H}_{M.t} = \vec{E}_{M.n} = \vec{H}_{J.n} = 0 \tag{196}$$

• The Huygens problem: With the sources $\Sigma(J, M)$, situated in Ω^-, are associated on S the secondary sources (123). These create in Ω^+ the same field as Σ and in Ω^-, a zero field:

$$\forall x \in \Omega^+ : \vec{E}_J(\underline{x}) + \vec{E}_M(\underline{x}) = \underline{\vec{E}_J}(x) - \underline{\vec{E}_M}(x) = 0 \tag{197}$$

From which, at x:

$$\vec{E}_J(x) = \vec{E}_M(x) = \frac{1}{2}\vec{E}(x) \tag{198}$$

In the same way:

$$\vec{H}_J(x) = \vec{H}_M(x) = \frac{1}{2}\vec{H}(x) \tag{199}$$

Each of the electrical and magnetic sources distributed in S creates half of the e.m. field in Ω^+ (the symmetries of the partial fields are neutralized in Ω^-).

In Ω^+ the total field is therefore radiated by the current $2\vec{J}_S = 2\vec{n} \wedge \vec{H}_S$ or else by the current $2\vec{M}_S = -2\vec{M} \wedge \vec{E}_S$. It is thus determined by the tangential component of the single electric field on S or of the single magnetic field (cf Dirichlet's problem).

• The reflector problem (see page 90): It is assumed that Ω^- is occupied by a perfect conductor and that the sources $\Sigma(J(x'), M(x'))$ are situated in Ω^+. The sources radiate in the empty space $\mathbf{R}^3$ a field $E^{(i)}(x)$, $H^{(i)}(x)$ (known as the incident field), which obeys the radiation conditions (see page 39). They radiate, in the presence of the conductor, a field E, H which obeys the boundary conditions on S and the conditions at infinity. This induces in S surface currents which, unlike the Huygens case, have a physical significance.

The extension to e.m. of the image method provides a simple solution, which is very useful in the study of antennas. The source images in Ω^-, $\Sigma'(-J(x'), M(x'))$, $x' \in \Omega^-$, are associated with Σ. These create, in Ω^+, the 'reflected' field:

$$E^{(r)}(x) = -\underline{E}^{(i)}(\underline{x}), \quad H^{(r)}(x) = \underline{H}^{(i)}(\underline{x}) \tag{200}$$

Hence, $\forall x \in \Omega^+$ the total field:

$$E(x) = E^{(i)}(x) + E^{(r)}(x) = E^{(i)}(x) - E^{(i)}(x) \tag{201}$$

$$H(x) = H^{(i)}(x) + H^{(r)}(x) = H^{(i)}(x) + H^{(i)}(x) \tag{202}$$

which obeys the boundary conditions (118) and, at infinity, the radiation conditions on S. In Ω^+, therefore, it coincides with the desired field.

Note

• Σ' can be replaced by the Huygens sources

$$\begin{aligned} \vec{J}'_S &= \vec{n} \wedge \vec{H}_S^{(r)} = \vec{n} \wedge \underline{\vec{H}_S^{(i)}} \\ \vec{M}'_S &= -\vec{n} \wedge \vec{E}^{(r)} = \vec{n} \wedge \underline{\vec{E}}^{(i)} \end{aligned} \tag{203}$$

and hence, in Ω^+, the field E, H by the combined radiation of Σ, J'_S, M'_S or $\Sigma, 2J'_S$ or $\Sigma, 2M'_S$.

• On S, the total induced currents are:

$$\begin{aligned} \vec{J}_S + \vec{J}'_S &= \vec{n} \wedge (\vec{H}_S^{(i)} + \underline{\vec{H}_S^{(i)}}) = 2\vec{n} \wedge \vec{H}^{(i)} \\ \vec{M}_S \rightarrow \vec{M}'_S &= -\vec{n} \wedge (E_S^{(i)} - \underline{E_S^{(i)}}) = 0 \end{aligned} \tag{204}$$

5: Far fields

(a) Formulae for far field radiation

The fact that a transmitting antenna has as its essential aim the production of a signal over a long distance justifies the stress on the asymptotic behaviour of the radiation fields.

The case of the dipole (177) has already been mentioned. The convergence theorem, summarized on page 31, allows the field radiated by a generalized antenna with a density of $\vec{K}(x')$ to be expressed in the form:

$$\vec{C}(x) = \int \vec{C}_0[x - x'; \vec{K}(x')]\, dx' \tag{205}$$

(177) can be used for this, taking into account the fact that $\vec{u}' = \vec{u} + o\left(\frac{1}{r}\right)$

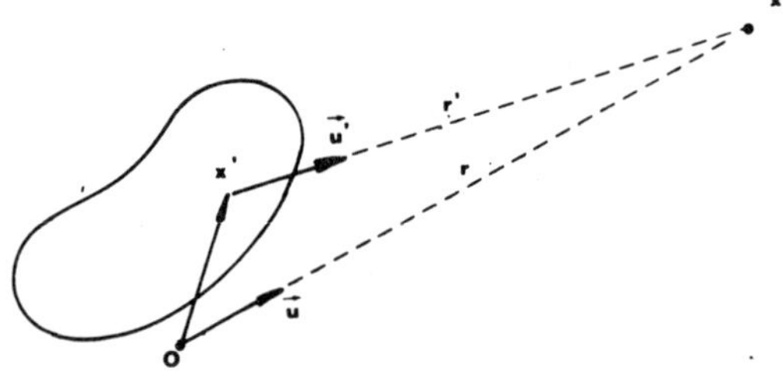

Fig. 4. *Antenna radiation.*

to obtain at long distance ($kr \gg 1$):

$$\vec{C}(x) = -ik \int [\vec{K}(x') \wedge u + j\vec{K}(x')] \wedge \vec{u} G(x - x')\, dx' + o\left(\frac{1}{r}\right) \tag{206}$$

in which the electric and magnetic parts can be separated:

$$\begin{aligned} \vec{E}(x) &= -ik \int [\eta(\vec{J} \wedge \vec{u}) \wedge \vec{u} - \vec{M} \wedge \vec{u}] G\, dx' + o\left(\frac{1}{r}\right) \\ \vec{H}(x) &= -ik \int \left[\vec{J} \wedge \vec{u} + \frac{1}{\eta}(\vec{M} \wedge \vec{u}) \wedge \vec{u}\right] G\, dx' + o\left(\frac{1}{r}\right) \end{aligned} \tag{207}$$

As shown above, the transposition of these formulae to the surface or linear distributions is straightforward.

(b) Simple antennas in a vacuum ($\vec{M} = 0, \eta = 120\pi$)
It is found immediately, taking account of:

$$G = \frac{e^{-ikr'}}{4\pi r'} = \frac{e^{-ikr}}{r} e^{ik(\vec{x}' \cdot \vec{u})} + o\left(\frac{1}{r}\right) \tag{208}$$

that the basic formulae of long-distance radiation for simple antennas are as below:

$$\vec{E} = 30ik \int (\vec{J} \wedge \vec{u}) \wedge \vec{u} \frac{e^{-ikr'}}{r'}\, dx' + o\left(\frac{1}{r}\right) \tag{209}$$

$$= 30ik \frac{e^{-ikr}}{r} \int (\vec{J} \wedge \vec{u}) \wedge \vec{u}\, e^{i\vec{k} \cdot \vec{x}'}\, dx' + o\left(\frac{1}{r}\right) \tag{210}$$

$$\vec{H} = \frac{ik}{4\pi} \int (\vec{J} \wedge \vec{u}) \frac{e^{-ikr'}}{r'} dx' + o\left(\frac{1}{r}\right) \tag{211}$$

$$= \frac{ik}{4\pi} \frac{e^{-ikr}}{r} \int (\vec{J} \wedge \vec{u}) \, e^{i\vec{k} \cdot \vec{x}'} dx' + o\left(\frac{1}{r}\right) \tag{212}$$

with the propagation vector $\vec{k} = k\vec{u}$.

(c) Radiating apertures

In Kirchoff's approximation, the formulae (208) allow the calculation at infinity of radiation from apertures. The surface integrals are simply extended to such an aperture with the surface currents given by (123).

(d) Characteristics of radiation

The electric field (209) is in the form:

$$\vec{E} = \vec{F}(\vec{u}) \frac{e^{-ikr}}{r} + o\left(\frac{1}{r}\right) \tag{213}$$

It is related to the magnetic field by:

$$\vec{E} - \eta \vec{H} \wedge \vec{u} = o\left(\frac{1}{r}\right) \tag{214}$$

equivalent to:

$$\vec{E} \wedge \vec{u} + \eta \vec{H} = o\left(\frac{1}{r}\right) \tag{215}$$

$\vec{F}(\vec{u}) = \vec{F}(\theta, \varphi)$ is a vector characteristic of radiation. It will be noted that:

$$\vec{E} = O\left(\frac{1}{r}\right) \tag{216}$$

$$\vec{H} = O\left(\frac{1}{r}\right) \tag{217}$$

The practical interest of these results (locally plane wave to first order) will become clear later on (Part 2). On a theoretical level it can be seen that the field at infinity of any limited source remains in the form given in (213). The following relationships are conditions of radiation, which are important in that they ensure the uniqueness of the solutions to any external problem.

6: Supplementary information

(a) Elementary solution to Helmholtz's vector equation

The transposition of (135) is not evident. Most English-speaking authors use Gibbs' dyadic calculation, which is simply a second-order tensor calculation. It is perhaps easier to reduce the problem to a particular convolution system (see

Appendix 3). Instead of (127), consider the matrix equation:

$$(\mathbf{A}) * \begin{pmatrix} C_1 \\ C_2 \\ C_3 \end{pmatrix} = \begin{pmatrix} S_1 \\ S_2 \\ S_3 \end{pmatrix} \tag{218}$$

where:

$$(\mathbf{A}) = \begin{pmatrix} (\Delta + k^2)\delta & 0 & 0 \\ 0 & (\Delta + k^2)\delta & 0 \\ 0 & 0 & (\Delta + k^2)\delta \end{pmatrix} \tag{219}$$

equivalent to:

$$\Delta C_i + k^2 C_i = S_i \tag{220}$$

An elementary solution is provided by the matrix:

$$(E) = \begin{pmatrix} G & 0 & 0 \\ 0 & G & 0 \\ 0 & 0 & G \end{pmatrix} \tag{221}$$

$$(\mathbf{A}) * (E) = \begin{pmatrix} \delta & 0 & 0 \\ 0 & \delta & 0 \\ 0 & 0 & \delta \end{pmatrix} = I_{3\times 3}\delta \tag{222}$$

(b) Conditions of radiation

These are boundary conditions at infinity, and useful uniqueness criteria for the solutions to external problems. Their importance was first shown by Sommerfeld. They have been verified previously, with particular solutions, but only a few of the results will be considered here. As already stated the trivial of the solutions to homogeneous equations should be emphasized.

(1) Homogeneous Helmhotz equation Let $U(x)$ be a solution to the equation:

$$\Delta U + k^2 U = 0 \tag{223}$$

in the exterior region Ω of a regular closed surface S, possibly whole space R^3.

The condition:

$$\partial_r U + ikU = o\left(\frac{1}{r}\right), \quad r \to \infty \tag{224}$$

(uniformly relative to the direction of x) is Sommerfeld's radiation condition (cf (140)). The solutions that obey it will be known as 'radiated'. The condition corresponds to the physical image of waves propagating towards infinity. It excludes waves 'originating' at infinity such as those which, inside a cavity, contribute to the formation of stationary waves.

Although weaker, Magnus's radiation condition can also be used:

$$\int_{|x|=r} |\partial_r U + ikU|^2 \, ds = o(1) \tag{225}$$

Kirchhoff's representation (Green's formula) is used to establish that the only radiated solution in the entire space is the trivial solution.

More generally, there is still a trivial solution in Ω if U or its normal derivative cancel themselves out on S. Finally, it can be shown that for any radiated solution in an external domain Ω, there exists a continuous function $F(\vec{u})$ on the unit sphere $|x| = 1$ such that:

$$U(x) = \frac{e^{-ikr}}{r} F(\vec{u}) + o\left(\frac{1}{r}\right) \tag{226}$$

Note

In the plane $\mathbf{R}^2$, the radiation condition becomes:

$$\partial_r U + ikU = o\left(\frac{1}{\sqrt{r}}\right) \tag{227}$$

(224) and (227) are simply specific examples of the general (though academic) case of $\mathbf{R}^n$:

$$\partial_r U + ikU = o\left(\frac{1}{r^{(n-1)/2}}\right) \tag{228}$$

(2) Maxwell's equations In an external region Ω where there are no sources (homogeneous Maxwell's equations), the fields E and H which satisfy (216), (217) and (214) (or 215) will be called the radiated solutions. It can be shown that the only radiated solution in the entire space ($\Omega = \mathbf{R}^3$) is the trivial solution:

$$\vec{E} = \vec{H} = 0$$

Hence the uniqueness of the various expressions for fields radiated by limited sources (such as (168), (169), (206) to (212)), all deduced from the fundamental result (146).

The analyticity of the fields outside the sources can be established, and from that it can be deduced that the only radiated solution in Ω such that on S

$$\vec{n} \wedge \vec{E} = \vec{n} \wedge \vec{H} = 0 \tag{229}$$

(zero tangential components) is the trivial solution. The conditions only need be obeyed on a regular part of S.

Finally, in parallel with the result in the previous paragraph, it can be shown that on the unit sphere $r = 1$ there is a continuous tangential vector $F(u)$ such that:

$$\vec{E} = \frac{e^{-ikr}}{r}\vec{F}(\vec{u}) + o\left(\frac{1}{r}\right)$$
$$\vec{H} = \frac{1}{\eta}\frac{e^{-ikr}}{r}\vec{u} \wedge \vec{F}(\vec{u}) + o\left(\frac{1}{r}\right) \tag{230}$$

hence the extreme generality of the idea of radiation characteristics.

(3) **Regularity of the surface** S The radiation conditions can cease to be uniqueness conditions in problems concerning non-regular boundaries (ridges, vertices, etc). The following is a simple example which illustrates this. Consider the diffraction of a plane wave by a perfectly conducting half-plane ($\theta = 0$ and 2π in semipolar coordinates (r, θ, z)). The problem was solved by Sommerfeld (see Chapter 3). The case will be limited here to a consideration of the situation in which the electric field is reduced to its component E_z parallel to the edge of the obstacle. This comes down to a two-dimensional scalar problem. Sommerfeld's solution is zero for each of the faces of the conductor, and is satisfied by (227). It is easy, however, to verify that there are an infinite number of solutions to Maxwell's equations with the diffracted part:

$$E_z = A_n H_n^{(2)}(kr) \sin n\theta \tag{231}$$

and satisfying the same conditions. Meixner's condition must be added (page 22), as it limits the rate of increase of the field around the edge, so as to isolate Sommerfeld's solution (the field (231) increases as $1/r^n$).

Chapter 2

Waves and Rays

A: Propagation of electromagnetic waves

1: Electromagnetic waves

The idea of waves is familiar to the physicist. It suggests the propagation of various quantities. This image is, however, quite inadequate and it is preferable to consider electromagnetic waves as signifying any solution to Maxwell's equations, in time and space.

The previous chapter was devoted to the study of fields as a function of the sources that create them. This chapter will consider their structure in external problems, that is, those to which the solutions are those of the homogeneous Maxwell's equations.

For sinusoidal fields, the phase value plays a major role, hence the importance of phase waves.

2: Homogeneous plane waves

These are the simplest solution to the homogeneous Maxwell's equations in a perfect medium (e, μ). They are sufficiently well known not to require elaborate explanation, but too important not to be briefly summarized here, at least from the point of view of their structure. Homogeneous plane waves are solutions in the form:

$$\begin{aligned}\vec{E}(x) &= \vec{E}_0\, e^{-i\vec{k}\cdot\vec{x}} = \vec{E}_0\, e^{-i(k_1x_1+k_2x_2+k_3x_3)}\\ \vec{H}(x) &= \vec{H}_0\, e^{-i\vec{k}\cdot\vec{x}}\end{aligned} \tag{1}$$

$\vec{k} = k\vec{u}$ is the propagation vector in the direction of the unit vector $\vec{u}$ and $k = \omega\sqrt{\varepsilon\mu}$ is the propagation constant in the medium.

The complex vectors $\vec{E}_0$ and $\vec{H}_0$ are constant, orthogonal to each other and the direction of propagation, so: $\vec{E}_0 = \eta(\vec{H}_0 \wedge \vec{u})$ where $\eta = \sqrt{\mu/\varepsilon}$.

Finally, a coordinate z in the direction of propagation will allow the propagation factor $\exp(-ikz)$ to be expressed. The description of the phenomena is immediate.

- The instantaneous relationships $\vec{\mathscr{E}}(x, t) = \eta\vec{\mathscr{H}}(x, t) \wedge \vec{u}$ expresses the fact that the triples $(\mathscr{E}, \mathscr{H}, u)$ is rectangular and right-handed (when taken in alphabetical order), and that the ratio of measurements of the two fields, at each point and at every instant, is the constant η, homogeneous at one impedance (that of the wave).

• The planes z = constant are of equal phase $\varphi(z, t) = \omega t - kz$ + constant for each of the components. The phase is propagated in the z direction increasing with the phase velocity: $v = \omega/k = \omega\sqrt{\varepsilon\mu}$ ($=c$ *in vacuo*).

The planes z = constant are said to be phase planes or wave planes, and the phase is said to propagate by plane waves in the direction of the propagation vector.

• The plane waves are known as homogeneous because of the equality of the ellipses (E) and of the ellipses (H).

• The phase is mod 2π for any displacement mod λ where: $\lambda = 2\pi/k$ is the wave length. The relationship *in vacuo*, λ (metres) f (megahertz) = 300, familiar to radio engineers, is easy to remember.

• The waves transport the power density: $P = E'^2/2\eta$.

Notes

• Homogeneous plane waves have no physical existence (they transport an infinite power!). They are non-trivial solutions to the homogeneous equations and do not satisfy the (partial) conditions of radiation (equation (191) of Chapter I). They are, nonetheless, very useful. In many problems they provide a good local representation of the true situation (far-field radiation, TEM waves, optics, etc). On a theoretical level, they can be used for such manipulations as diffraction of a plane wave, or Fourier-type syntheses.

• In geometrical optics, propagation is seen as the transport of power along rays orthogonal to the wave planes. The optical path takes the role of the phase: $(k/k_0)z$ (where $k = k_0$ *in vacuo*).

3: Various extensions

(a) Conductive media

A complex constant can be used:

$$k = k' - ik'' \tag{2}$$

The propagation factor

$$e^{-ikz} = e^{-k''s}\, e^{-ik'z} \tag{3}$$

reveals an attentuation constant k''.

The complex vectors E and H are, once again, orthogonal, but this is not so for the real fields they represent, despite common belief to the contrary.

Note that a complex propagation vector can be defined:

$$\vec{k}_c = k'\vec{u} - ik''\vec{u} = \vec{k}' - i\vec{k}'' \tag{4}$$

(b) Dispersive media

When the frequency varies and the relationship between ω and $k(\omega)$ is no longer a direct ratio, waves of different frequencies have different phase velocities:

$$v(\omega) = \frac{\omega}{k(\omega)} \tag{5}$$

Propagation disperses them. Consider a situation in which the waves interfere in a narrow band around an angular frequency ω (group or packet of waves).

At fixed z_0 and t_0, the following phases exist: $\varphi(\omega) = \omega t_0 - k(\omega) z_0$. The amplitude of each field component is maximal for the waves 'in phase', that is, obeying $d\varphi = t_0\, d\omega - z_0\, dk = 0$.

The link thus established between z_0 and t_0 defines the velocity of the maximum amplitude or the group velocity:

$$v_g = \frac{d\omega}{dk} = \frac{1}{k'(\omega)} \tag{6}$$

This elegant reasoning is due to Born.

(c) Complex plane waves

1. Definition Conductive media have required, above, the use of a complex propagation vector:

$$\vec{k} = \vec{k}' + i\vec{k}'' \tag{7}$$

the real and imaginary parts of which were parallel. Undamped propagation is the limiting case: $k'' = 0$. For reasons of symmetry, the limiting case $k' = 0$ could be envisaged. This is simply the case of evanescent waves where, for each component, there is an oscillation in phase, with amplitude decreasing exponentially in the direction of $-k''$.

The idea of an extension, if only formal, of (7) to any complex vector is quite natural, and has useful applications. A homogeneous plane wave

$$e^{-i\vec{k}\cdot\vec{x}} = e^{\vec{k}''\cdot\vec{x}}\, e^{-i\vec{k}'\cdot\vec{x}} \tag{8}$$

will be said to be complex (not to be confused with the complex representation of the field transported by an ordinary plane wave). It raises no problem with calculation, which remains the same, but must be interpreted.

2. Mathematical aspect For the purposes of this discussion, only a perfect medium will be considered here. Thus:

$$|\vec{k}|^2 = |\vec{k}'|^2 - |\vec{k}''|^2 - 2i\vec{k}'\cdot\vec{k}'' \quad \text{or} \quad k^2 = k'^2 - k''^2 - 2i\vec{k}'\cdot\vec{k}'' \tag{9}$$

is real, and so:

$$k'^2 - k''^2 = k^2 \tag{10}$$

$$\vec{k}'\cdot\vec{k}'' = 0 \tag{11}$$

The real and imaginary parts of the propagation vector are orthogonal. Their plane is taken as (x_1, x_2) of a cartesian coordinate system (x_1, x_2, x_3). The electrical (or magnetic) field can be assumed to be polarized along Ox_3.

The components of the propagation vector remain in the following form:

$$\begin{aligned} k_1 &= k_1' + ik_1'' = k\cos\theta \\ k_2 &= k_2' + ik_2'' = k\sin\theta \end{aligned} \tag{12}$$

on condition that $\theta = (Ox_1, \vec{k}) = \theta' + i\theta''$ is complex, such that:

$$k_1 = k(\cosh\theta'' \cos\theta' - i \sinh\theta'' \sin\theta') \tag{13}$$

$$k_2 = k(\cosh\theta'' \sin\theta' + i \sinh\theta'' \cos\theta') \tag{14}$$

thus obeying (10) and (11).

The exponential (8) will represent a homogeneous plane wave, propagating in an imaginary direction defined by the complex angle of the propagation vector with the axis x_1.

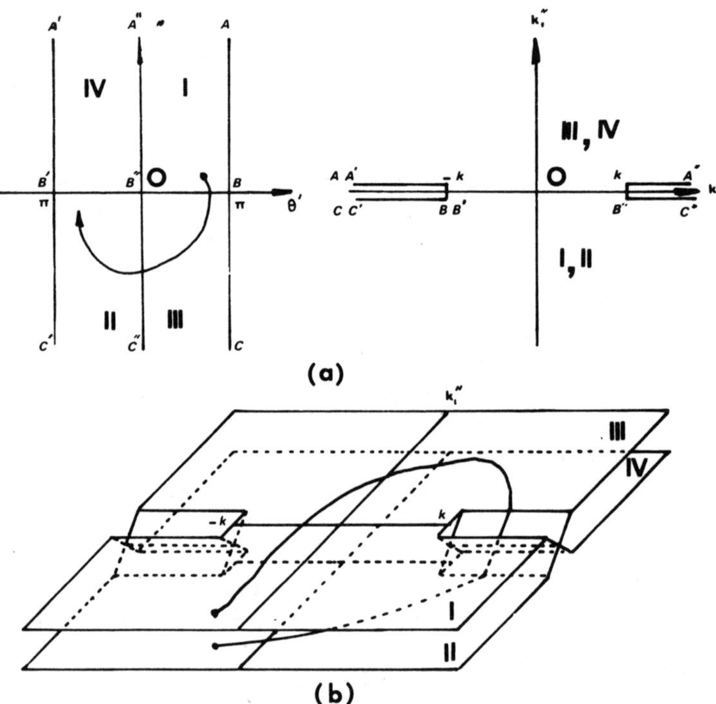

Fig. 5. *Complex plane waves.*

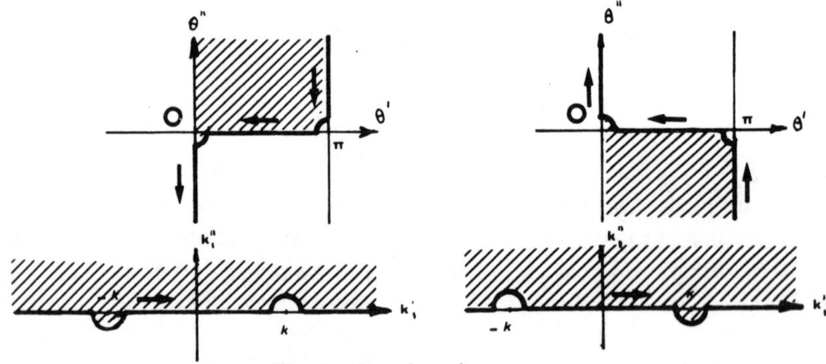

Fig. 6. *Complex plane waves.*

This concept is very important. The relationship $k_1^2 + k_2^2 = k^2$, which in the case of a real direction $\theta'' = 0$ implies $|k_1|, |k_2| \leqslant k$, can be extended to any real k_1 and k_2 in the directions $\theta = i\theta''$, where:

$$k_1 = k \cosh \theta'' > k \quad , \quad k_2 = k \sinh \theta'' \tag{15}$$

and in the directions $\theta = \pi + i\theta''$ where:

$$k_1 = -k \cosh \theta'' < -k \quad , \quad k_2 = -k \sinh \theta'' \tag{16}$$

This is the key to using the Fourier Transform (see below) in radiation problems. It allows integration over the whole plane (k_1, k_2).

More generally, complex plane waves offer a great deal of flexibility in the representation of plane fields by integrals; for example:

$$E(\vec{x}) = \int_\Gamma A(\theta)\, e^{-i\vec{k}\cdot\vec{x}}\, d\theta \tag{17}$$

along the appropriate arcs Γ of the complex plane θ (leady waves, plane diffraction, etc).

The following points should be noted. Equation (13) defines the mapping

$$k_1' = k \cosh \theta'' \cos \theta' \quad , \quad k_1'' = -k \sinh \theta'' \sin \theta' \tag{18}$$

of plane (θ', θ'') in the plane (k_1', k_1'').

Each of the half-strips I and II (Figure 5a) (resp. III and IV) corresponds bijectively to the half-plane $k_1'' < 0$ (resp. > 0). The band period $-\pi \leqslant \theta' \leqslant \pi$ (or more generally: $\theta_0' \leqslant \theta' \leqslant \theta_0' + 2\pi$) is therefore applied bijectively to a Riemann surface with two sheets communicating through the cuts $|k_1'| \geqslant k$, $k_1'' = 0$ (Figure 5b).

Finally, the two contours C and C' will have to be used (Figure 6), corresponding to real k_1 and to the respective radiation condition at infinity in the half planes $x_2 \lessgtr 0$.

3. Physical aspect Relationship (8) reveals the two orthogonal families of planes:

$$\vec{k}'\cdot\vec{x} = \text{constant} \quad , \quad \vec{k}''\cdot\vec{x} = \text{constant} \tag{19}$$

The first are phase planes, and propagate in the direction of vector $\vec{k}'$ of length:

$$k' = k \cosh \theta'' > k$$

and inclined by θ' to Ox_1. Their velocity is:

$$\frac{\omega}{k'} = \frac{\omega}{k \cosh \theta''} = \frac{v}{\cosh \theta''} < v \tag{20}$$

The second are amplitude planes, and decrease exponentially towards O in the direction $-k''$ perpendicular to that of propagation.

This particular inhomogeneous plane wave structure is that of the waves said to be dissociated. It provides the physical interpretation of the complex homogeneous plane waves that were introduced as a mathematical device.

The behaviour of the amplitude in radiation problems means that regions in which this amplitude would become infinite at infinity (improper waves) must be rejected.

Note that, in (13) and (14), the vector $\vec{k}''$ of length $k|\sinh \theta''|$ is inclined to Ox_1 by $\theta' \pm \pi/2$, $\pm$ according to the sign of θ''.

4. Applications

• Radiation in the half-planes $x_2 > 0$ and $x_2 < 0$. Taking account of the above, real k_1, k_2 are ensured by:

1 $\theta = \theta'$: propagation in all real directions or
2 $\theta = i\theta''$: $\theta'' > 0$ (resp. < 0): $-k''$ directed towards $x_2 < 0$ (resp. > 0) or
3 $\theta = \pi + i\theta''$: $\theta'' > 0$ (resp. < 0): $-k''$ directed towards $x_2 > 0$ (resp. < 0).

Hence the two contours described in the complex plane θ (Figure 6), relative respectively to the damped propagation of the waves on side $x_2 > 0$ or on the side $x_2 < 0$, and to that of the homogeneous waves in the whole plane.

• Complex rays: Geometrical optics (see below) substitutes rays for the plane waves. A ray is any direction aligned with the propagation vector. It is helpful to extend the concept to the case of complex waves. The great advantage of complex rays is that it is possible to take account of evanescent-type fields, retaining the language of optical geometry, whereas this is not normally the case (total reflection, neighbourhood of caustics, etc).

Example

Refraction law: Consider, in a half-plane $x_2 < 0$, a ray inclined by θ to Ox_1. This gives rise to a reflected ray and a refracted ray inclined by θ_r in the half-plane $x_2 > 0$, such that: $\cos \theta_r = (\cos \theta_i)/n$. When $n < 1$, there is total internal reflection if $\theta_i < \text{arc} \cos n$. In geometric optics, there is no penetration into the second medium, which is experimentally not the case. It is possible to extend the above relationship to $\theta_r = i\theta_r''$, $\theta_r'' < 0$, to take account of the parallel propagation to Ox_1 of a dissociated wave with amplitude decreasing exponentially towards $x_2 > 0$.

• Surface waves: Vol. 2 of this work contains a discussion of the open guiding structures to which the above considerations apply, at least in plane models.

Consider the half-space $x_2 \leqslant 0$, occupied by a physical medium, natural or otherwise, which guides a dissociated wave parallel to the axis Ox_1, in the half space $x_2 \geqslant 0$. The values $\theta' = 0$, $\theta'' < 0$ can be adopted.

This gives the image of a wave propagating without damping, parallel to Ox_1, with a velocity (20) less than that of the free waves in the medium $x_2 > 0$ (slow medium).

The amplitude, constant on the planes $x_2 = $ constant, decreases exponentially further from the line. Concentration of the fields about the line can be characterized by the attenuation constant $\alpha = -k \sinh \theta''$.

• Leaky wavcs: Surfacc waves which are undamped in the direction of the guide do not give rise to a lateral radiation of energy (see below). There are, however, waveguide structures with which, on the contrary, this type of radiation is associated. The theoy of these leaky waves is based on a tricky discussion of representation (17), and will not be described here. The following basic points may, however, be useful.

Consider the complex wave (8) at a point $x = (r, \varphi)$ in the real direction φ, the propagation vector having the complex direction θ. This gives the polar representation:

$$E = e^{-kr \sin(\theta' - \varphi) \sinh \theta''} \cdot e^{-ikr \cos(\theta' - \varphi) \cosh \theta''} = A\, e^{-ik_\varphi r} \qquad (21)$$

Phase: With each direction φ, there can be associated a phase velocity (for example, *in vacuo*) $v_\varphi = c/\cos(\theta' - \varphi) \cosh \theta''$, which is minimum in the direction of the vectors $\pm k'(\varphi = \theta', \bmod \pi)$ with the value $c/\cosh \theta'' < c$ found for the surface waves. Propagation remains 'slow' ($v_\varphi < c$) for angles $\theta' - \alpha < \varphi < \theta' + \alpha$, where $\cos \alpha = 1/\cosh \theta''$. It is 'rapid' ($v_\varphi > c$) outside that angle. The phase velocity is even infinite in the directions $\pm k''$.

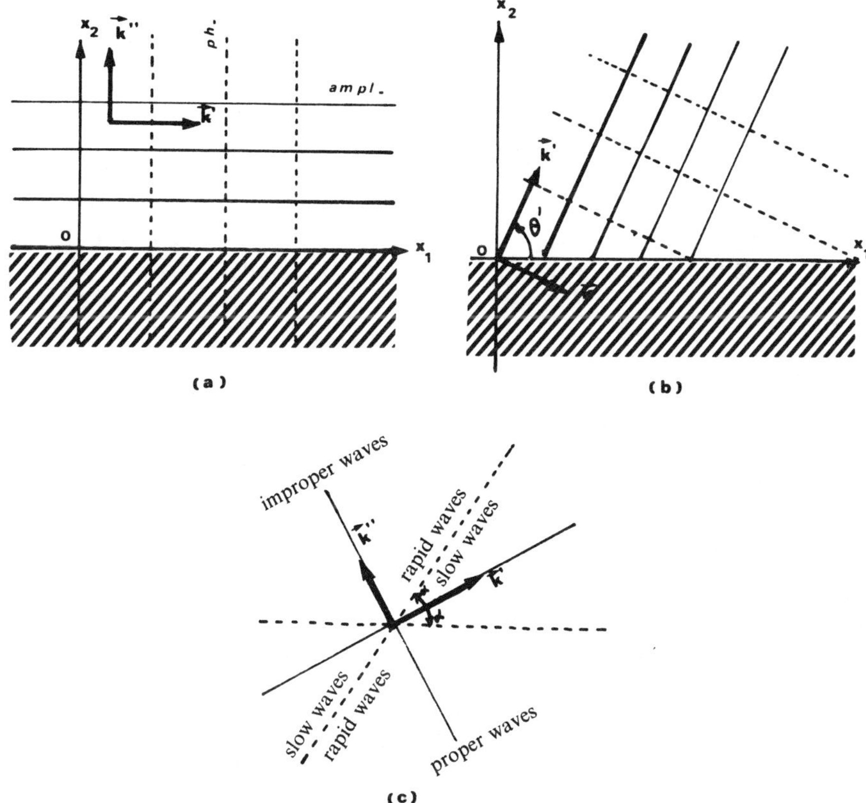

Fig. 7. *Surface waves and leaky waves.*

Amplitude: This tends exponentially towards 0 for $r \to \infty$ when $\sin(\theta' - \varphi) \cdot \sinh \theta'' > 0$; that is, on the side $-k''$ (proper wave idea).

Figure 7(b) corresponds to a technical situation with the improper wave on the side $x_2 > 0$, which has physical meaning in an appropriate domain only. If, for example, the excitation takes place at O, the field is considered only in sector $(Ox_1, \vec{k}')$.

Along $x_2 =$ constant, parallel to the structure, the fields vary as follows:

$$e^{-kx_1 \sinh\theta'' \sin\theta'} \, e^{-ikx_1 \cosh\theta'' \cos\theta'}$$

There is damped propagation with phase velocity $v/\cosh\theta'' \cos\theta'$ greater than v (rapid wave) if $\cosh\theta'' \cos\theta' < 1$. This is encountered in a number of problems: radiation from the simple guide, with slots or apertures, periodic structures with a large period, log-periodic antennas, Wood anomalies, Cherenkov radiation, etc.

Notes

• The mathematical model is only an intermediary which represents improperly (improper wave) the physical phenomena. For fixed x_1 there would be an amplitude increasing indefinitely with x_2 (Figure 7(b)).

• The electric field is assumed to be polarized along Ox_3, with an amplitude E at the origin, so the calculation of the magnetic field, starting with Poynting's vector, is straightforward. It is easy to find:

$$P = \frac{E^2}{2\eta} e^{-2\vec{k}'' \cdot \vec{x}} \vec{k}^* \tag{22}$$

the real part of which, $(E^2/2\eta)\exp(-2\vec{k}'' \cdot \vec{x})\vec{k}'$, allows it to be verified that a surface wave transports the active energy along the guiding structure. There is no lateral radiation. A leaky wave, however, gives rise to continuous oblique radiation. This is why the wave is damped in the direction of the structure.

(d) Locally plane waves

Plane TEM electromagnetic waves, as used in lines, are not homogeneous but obey (at every point) the same field relationships as the homogeneous plane waves. The previous chapter contained expressions (230) for fields radiated at long distance by finite sources. If the radiation characteristic F is assumed to be real, it can be said that the principal parts of the fields propagate by spherical phase waves. In addition, at each point of a phase sphere a tangential homogeneous plane wave structure can be recognized. A receiving antenna—a finite, diffracting obstacle—in the far field of the sources will therefore receive a portion of a homogeneous plane wave.

Two-dimensional problems present a similar situation to that of cylindrical phase waves. The approximation of geometrical optics provides another example. It uses any wave surfaces, each little part of which can be considered as an element of a homogeneous plane wave.

These diverse examples make the concept of the locally plane wave much easier to grasp.

4: The electromagnetic field and the Fourier Transform (FT)

(a) Role of the Fourier Transform

The FT was discussed in relation to the complex representation of the field (page 8). At a fixed point x it substitutes its harmonic representation, for example at frequency ω, for a field expressed as a function of time t. This allows the simplification brought about by the sinusoidal signals to be exploited in the domain.

Consider a complex field. Note that the manipulation of plane homogeneous waves owes its convenience to the fact that they depend sinusoidally on the space variables. As a result, it is tempting to introduce an FT relative to the coordinates x_i and components k_i. This technique is very useful in the study of antennas and has become widespread.

(b) Superposition of plane waves

In a perfect medium (ε, μ) and for sinusoidal fields (ω) such that

$$|\vec{k}|^2 = k^2 = k_1^2 + k_2^2 + k_3^2 = \omega^2 \varepsilon \mu = \text{constant} \tag{23}$$

the homogeneous Maxwell equations have solutions in the form

$$\begin{aligned} \vec{E}(x) &= \vec{E}_0\, e^{-i\vec{k}\cdot\vec{x}} (\vec{E}_0 \perp \vec{k}) \\ \vec{H}(x) &= \frac{1}{\omega\mu} \vec{k} \wedge \vec{E}_0\, e^{-i\vec{k}\cdot\vec{x}} \end{aligned} \tag{24}$$

and the same applies for any finite linear combination of them, and by continuity, any limit of sequences of such combinations.

Notation

For the sake of simplicity, the following notation will be used:

- Observation point

$$\vec{r} = (x, y, z), \quad \vec{\rho} = \frac{\vec{r}}{\lambda} = (\xi, \eta, \zeta) \quad , \quad \xi = \frac{x}{\lambda}, \ldots \tag{25}$$

- Propagation vector

$$\begin{aligned} \vec{k} &= k\vec{u} = (k\alpha, k\beta, k\gamma) \\ |\vec{u}|^2 &= \alpha^2 + \beta^2 + \gamma^2 = 1 \end{aligned} \tag{26}$$

The interference of, for example, a continuous infinity of plane waves propagating in all directions $\vec{u}$ with the same phase constant k, creates fields that can be expressed:

$$\vec{E}(x, y, z) = \int_S \vec{A}(\alpha, \beta, \gamma)\, e^{-ik(\alpha x + \beta y + \gamma z)}\, dS \tag{27}$$

$$\vec{H}(x, y, z) = \frac{1}{\eta} \int_S \vec{u} \wedge \vec{A}\, e^{-ik(\alpha x + \beta y + \gamma z)}\, dS \tag{28}$$

or with reduced coordinates (25):

$$\vec{E}(\xi, \eta, \zeta) = \int_S \vec{A}(\alpha, \beta, \gamma)\, e^{-2\pi i(\alpha\xi + \beta\eta + \gamma\zeta)}\, dS \tag{29}$$

$$\vec{H}(\xi, \eta, \zeta) = \sqrt{\frac{\varepsilon}{\mu}} \int_S \vec{u} \wedge \vec{A}\, e^{-2\pi i(\alpha\xi + \beta\eta + \gamma\zeta)}\, ds \tag{30}$$

S is the unit sphere (26) in the space (α, β, γ). $\vec{A}\, dS$ plays the role of the excitation field $\vec{E}_0$ of (24). It represents the contribution of the 'small' packet of waves in directions close to each direction $\vec{u}$ and obeys:

$$\vec{A} \cdot \vec{u} = 0 \tag{31}$$

(c) Resolution into plane waves

The reverse problem arises naturally: that of resolving a given field into plane waves, that is, representing them in a form of the type (27), or determining the excitation characteristic $\vec{A}(\vec{u})$ in various directions starting from the field $\vec{E}(\vec{r})$. This would be a simple Fourier inversion if it were not for the problem posed by the wave number space being in only two dimensions (α, β, γ linked by (26)). A simple use of the normal Fourier Transform involves the artifice of fixing one of the coordinates, ζ for example. The wave number space and each of the planes are made to correspond: ζ = constant.

Note

The problem is easily solved if it is noted that the field is really a Fourier-Stieltjes transform, or the FT of a surface measure on the sphere S (see below).

(d) Two-dimensional problems

(1) Plane fields The particular case of fields depending only on x and y (or ξ and η) coordinates is suitable for the familiar types of calculation, the results of which are very useful (linear source models, separable plane apertures, etc). By adding the hypothesis that one of the fields, E for example, is polarized in the z-direction, the following simple scalar relationships can be found:

$$E(x, y) = \int_\Gamma A(\alpha, \beta)\, e^{-ik(\alpha x + \beta y)}\, ds \tag{32}$$

or

$$E(\xi, \eta) = \int_\Gamma A(\alpha, \beta)\, e^{-2\pi i(\alpha\xi + \beta\eta)}\, ds \tag{33}$$

(Γ: circle $\alpha^2 + \beta^2 = 1$)

(2) Polar representation In polar coordinates, (32) is:

$$E(r, \theta_0) = \int_C F(\theta)\, e^{-ikr\cos(\theta - \theta_0)}\, d\theta \tag{34}$$

whcrc $F(\theta) = A(\cos\theta, \sin\theta)$. The integral extended over any trajectory C in the complex plane $\theta = \theta' + i\theta''$ (see above) gives, formally at least, a great deal of flexibility in the representation of the electromagnetic fields. One of the trajectories on page 46 would, specifically, be suitable.

(3) **Far field** An interesting consequence of (34) is that $F(\theta)$ can be interpreted as a radiation characteristic. At long distances ($kr \gg 1$), the behaviour of the integral (34) is obtained by the stationary phase method (Appendix 2). The phase of the integrand is stationary for $\theta = \theta_0$. Formula (2) of the Appendix gives:

$$E(r, \theta_0) \simeq \sqrt{2\pi}\, e^{i(\pi/4)} F(\theta_0) \frac{e^{-ikr}}{\sqrt{kr}} \tag{35}$$

and hence the interpretation. The expression obtained obeys the radiation condition (Chapter 1, equation (127)). Note that formula (34) allows the field to be deduced at any point, starting from its values at infinity.

Example
Radially symmetric radiation: $F(\theta) = 1$

$$E(r) = \int_C e^{-ikr\cos\theta}\, d\theta \tag{36}$$

with, as contour C, that of Figure 6 rotated by $-\pi/2$. The integral is standard ([37], page 178):

$$E(r) = \pi H_0^{(2)}(kr) \tag{37}$$

This is, to the nearest factor, the simple solution (Chapter 1, equation (141)) to the Helmholtz equation in $\mathbf{R}^2$. As might be expected, the field is that of a uniform source concentrated on the z-axis.

Note

• (37) and (35) can be brought closer together to give the asymptotic expression of Hankel's function:

$$H_0^{(2)}(kr) \simeq \sqrt{\frac{2}{\pi}}\, e^{i(\pi/4)} \frac{e^{-ikr}}{\sqrt{kr}}$$

• The magnetic field $H_r = H_z = 0$, $H_\theta = iH_1^{(2)}(kr)/\eta$ is associated with the electric field $E_r = E_\theta = 0$, $E_z = H_0^{(2)}(kr)$.

(4) **Using the normal Fourier Transform** Consider (33) again, with α as the integration variable on the whole of the real axis. The condition $\alpha^2 + \beta^2 = 1$ associates a doubly determined function $\beta(\alpha)$ with α. In the calculation that follows, β will be the appropriate determination (see above) that provides the field, either in the half-plane $y > 0$, or in the half-plane, $y < 0$. Taking into account

$d\alpha = d(\cos\theta) = -\beta\, d\theta$, the following is found:

$$E(\xi, \eta) = -\int_{-\infty}^{\infty} \frac{A[\alpha, \beta(\alpha)]}{\beta(\alpha)}\, e^{-2\pi i\beta(\alpha)\eta}\, e^{-2\pi i\alpha\xi}\, d\alpha \qquad (38)$$

which expresses the fact that for each fixed η, the field varies with ξ with the normal Fourier Transform:

$$E(\xi, \eta) = \mathscr{F}\left\{-\frac{A}{\beta}\, e^{-2\pi i\beta\eta}\right\} \qquad (39)$$

This gives, if the inversion is valid:

$$A[\alpha, \beta(\alpha)] = -\beta(\alpha)\, e^{2\pi i\beta(\alpha)\eta} \int_{-\infty}^{\infty} E(\xi, \eta)\, e^{2\pi i\alpha\xi}\, d\xi \qquad (\eta = \text{constant}) \quad (40)$$

In particular, starting with the field on the ξ-axis:

$$A = -\beta(\alpha) \int_{-\infty}^{\infty} E(\xi, 0)\, e^{2\pi i\alpha\xi}\, d\xi \qquad (41)$$

Note

The analogy of expression suggests the language of signal theory, the relationship between time and frequency. There is no inconvenience involved in treating ξ as a (spatial) time, α as a (spatial) frequency, and A as a spectral representation. This is what optical scientists do. Specialists in antennas prefer to do the reverse.

• Application to antennas: Taking into account the previous paragraph, this very important result follows: there is a Fourier correspondence between linear sources and their radiation at infinity. If, conforming to the above note, the reduced abscissa on the source is treated as a frequency (spatial frequency), the direction cosine of the direction of radiation plays the role of a time. In fact, this use of language is very convenient, since it allows the results of signal theory (Vol. 2) to be taken over altogether. In particular, since any practical source is of finite extent, the radiation characteristics have the properties of limited spectrum signals (sampling theory, Bernstein's theorem, etc). It is interesting to underline the following: a function and its Fourier Transform cannot, at the same time, be of limited extent, since the radiation of limited sources must be interpreted for all real α, and consequently for complex directions (with non-visible regions).

• Huygens' principle: (41) determines $A(\alpha)$ on the basis of the repartition of the electric field on the ξ-axis. Hence, the field is known at any point on the half-place $\eta > 0$ (or $\eta < 0$) immediately, using (38). This application will be discussed further.

(e) Three-dimensional problems

The theoryis contained in formulae (27) to (30). Condition (31), where $\alpha A_\xi + \beta A_\eta + \gamma A_\zeta = 0$, determines the component A_ζ as a function of the two others.

Designation by $\vec{A}_t$ of the vector (A_ξ, A_η) immediately gives:

$$\vec{A} = \frac{1}{\gamma}(\vec{n} \wedge \vec{A}_t) \wedge \vec{u} \tag{42}$$

(1) Polar representation In polar coordinates, where:

$$\begin{aligned} \vec{u} &= (\sin\theta\cos\varphi, \sin\theta\sin\varphi, \cos\theta) \\ \vec{\rho} &= (\rho\sin\theta_0\cos\varphi_0, \rho\sin\theta_0\sin\varphi_0, \rho\cos\theta_0) \end{aligned} \tag{43}$$

(29) is expressed:

$$\vec{E}(\rho, A_0, \varphi_0) = \int_s \vec{A}(\theta, \varphi)\, e^{-2\pi i\rho\cos\psi} \sin\theta\, d\theta\, d\varphi \tag{44}$$

where ψ is the angle $(\vec{u}, \vec{\rho})$ with:

$$\cos\psi = \cos\theta\cos\theta_0 + \sin\theta\sin\theta_0\cos(\varphi - \varphi_0) \tag{45}$$

maximum in the direction: $\theta = \theta_0$, $\varphi = \varphi_0$.

The stationary phase method (Appendix 2, formula (7)) provides the asymptotic expression:

$$\vec{E}(\rho, \theta_0, \varphi_0) \simeq i\vec{A}(\theta_0, \varphi_0)\frac{e^{-2\pi i\rho}}{\rho} \tag{46}$$

or, returning to $r = \lambda\rho$:

$$\vec{E}(r, \theta_0, \varphi_0) \simeq 2\pi i\vec{A}(\theta_0, \varphi_0)\frac{e^{-ikr}}{kr} \tag{47}$$

$\vec{A}$ therefore has the significance of a radiation characteristic. The field in the space is, here also, determined by the field at infinity.

(2) Normal Fourier Transform The same method is used here as in point 3 above, and allows the normal Fourier Transform to be used in $\mathbf{R}^2$. Equation (26) is applied to associate $r(\alpha, \beta)$ with the independent variables α and β. If it is noted that $r\,dS = d\alpha\, d\beta$, (29) can be rewritten explicitly in the form:

$$\vec{E}(\xi, \eta, \zeta) = \int_{\mathbf{R}^2} \frac{\vec{A}[\alpha, \beta, \gamma(\alpha, \beta)]}{\gamma(\alpha, \beta)}\, e^{-2\pi i\gamma(\alpha,\beta)\zeta}\, e^{-2\pi i(\alpha\xi + \beta\eta)}\, d\alpha\, d\beta \tag{48}$$

which expresses, in each fixed plane ζ = constant, the distribution of the field at ξ, η and the normal Fourier Transform with two variables:

$$\vec{E}(\xi, \eta, \zeta) = \mathscr{F}\left\{\frac{\vec{A}}{\gamma}\, e^{-2\pi i\gamma\zeta}\right\} \tag{49}$$

The need to extend the integral (48) to the whole plane (α, β) means that one must distinguish the real directions $(\alpha^2 + \beta^2 \leqslant 1)$ from the complex directions $(\alpha^2 + \beta^2 > 1)$, that is, between two determinations of $\gamma = \cos\theta$, ensuring that the field remains finite in the half-space $\zeta > 0$ or the half-space $\zeta < 0$.

Exercise

Consider a real φ with $\theta = \theta' + i\theta''$. Verify the following results: The imaginary directions correspond to $\theta = (\pm\pi/2) + i\theta''$. The phase propagates in the horizontal direction φ with the constant $k \cosh \theta'' > k$ (slow wave). For $\theta' = \pi/2$ (resp. $-\pi/2$) the amplitude is zero at infinity on the side $\zeta > 0$ for $\theta'' > 0$ (resp. < 0). Define the corresponding contours of the complex plane θ and the determinations γ to be associated with them.

Formally, at least, the following is found by inversion of (49):

$$\vec{A} = \gamma e^{2\pi i \gamma \zeta} \mathscr{F}^{-1}\{\vec{\mathrm{E}}\} \tag{50}$$

(the Fourier Transform operates only on the variables ξ, η).

In particular, in the plane $\zeta = 0$:

$$\vec{E}(\xi, \eta, 0) = \mathscr{F}\left\{\frac{\vec{A}}{\gamma}\right\} \tag{51}$$

$$\vec{A} = \gamma \mathscr{F}^{-1}\{\vec{E}(\xi, \eta, 0)\} = \gamma \mathscr{F}^{-1}\{\vec{E}^{(0)}\} \tag{52}$$

giving the important result:

$$\vec{E}(\xi, \eta, \zeta) = \mathscr{F}\{e^{-2\pi i \gamma \zeta} \mathscr{F}^{-1}\{\vec{E}^{(0)}\}\} \tag{53}$$

which expresses the field in space given its values on the plane $\zeta = 0$ (or on any given plane). In fact, the field is known only in each half-space $\zeta \lessgtr 0$ according to the determination of γ.

It is even possible to start only with the 'tangential' component $\vec{E}_t^{(0)}$ in the plane $\zeta = 0$. By projecting (52) onto this plane, one obtains:

$$\vec{A}_t = \gamma \mathscr{F}^{-1}\{\vec{E}_t^{(0)}\} \tag{54}$$

which is sufficient to determine $\vec{A}$ by (42).

Note

(30) will give the magnetic field. It would have been possible to conduct the calculation, in an alternative manner and to determine the e.m. field from the 'tangential' magnetic field $H_t^{(0)}$ alone, on the plane $\zeta = 0$.

(f) Various applications

(1) Field at infinity of radiating apertures The model of the radiating aperture is encountered in many applications. An 'illumination' characteristic $\vec{E}_t^{(0)}(\xi, \eta)$ is taken for a limited domain Ω in plane $\zeta = 0$. From (54) and (42) it follows that:

$$\vec{A} = \int_\Omega (\vec{u} \wedge E_t^{(0)}) \wedge \vec{u}\, e^{2\pi i(\alpha\xi + \beta\eta)}\, d\xi\, d\eta \tag{55}$$

which constitutes a radiation characteristic in the fixed direction $\vec{u} = (\theta, \varphi)$. Thus, from (46) or (47):

$$\vec{E}(\rho, \vec{u}) = i\frac{e^{-2\pi i \rho}}{\rho} \int_\Omega (\vec{n} \wedge \vec{E}_t^{(0)}) \wedge \vec{u}\, e^{2\pi i(\alpha\xi + \beta\eta)}\, d\xi\, d\eta$$

or

$$\vec{E}(r \cdot \vec{u}) = -2ikG \int_{\Omega} (\vec{n} \wedge \vec{E}^{(0)}) \wedge \vec{u}\, e^{ik(\alpha x + \beta y)}\, dx\, dy \tag{56}$$

This would give a corresponding formulation of the magnetic field, from which could be deduced:

$$\vec{E} = \eta \vec{H} \wedge \vec{u} = 2ik\eta G\vec{u} \wedge \int_{\Omega} (\vec{u} \wedge \vec{H}^{(0)})\, e^{ik(\alpha x + \beta y)}\, dx\, dy \tag{57}$$

where $\eta = \sqrt{\mu/\varepsilon)}$

(2) Huygens' representation (p. 37) The importance of the previous formulae is that the electric and magnetic fields are expressed in the half-space $z > 0$; for example, starting with one given field, and even its tangential components, in the plane $z = 0$. These are the minimum data required for a Dirichlet-type problem.

The half-sum of the expressions (56) and (57) of the same electric field would give rise to the Huygens representation in the specific case of a plane boundary.

$$\vec{E} = ikG\vec{u} \wedge \int_{\Omega} [\vec{n} \wedge \vec{E}^{(0)} + \eta(\vec{n} \wedge \vec{H}^{(0)}) \wedge \vec{u}]\, e^{ik(\alpha x + \beta y)}\, dx\, dy \tag{58}$$

The comparison of (56) with (207) of Chapter 1 leads to the introduction of the magnetic surface current:

$$\vec{M}_s = -2\vec{n} \wedge \vec{E}^{(0)} \tag{59}$$

The physical interpretation of the coefficient is as follows. Suppose plane $z = 0$ is conductive so that, in conformity with Huygens' principle, the field is zero on the $z < 0$ side, and the magnetic current $-\vec{n} \wedge \vec{E}^{(0)}$ creates the discontinuity linked to the existence of the field on the $z > 0$ side. Thus there is a radiating source in the presence of a conducting plane.

The principle of images (page 90) is equivalent to making the double source radiate in the absence of the conductor. The same reasoning applies for the electric current associated with the magnetic field.

(3) Kirchhoff's representation The convolution theorem applied to (53) gives:

$$\vec{E} = \mathscr{F}\{e^{-2\pi i\gamma\zeta}\} * \mathscr{F}\{\mathscr{F}^{-1}\vec{E}^{(0)}\} = \vec{E}^{(0)} * \mathscr{F}\{e^{-2\pi i\gamma\zeta}\} \tag{60}$$

Exercise

Fourier transform of $e^{-2\pi i\gamma\zeta}$.

The Fourier transform of $e^{-2\pi i\gamma\zeta}/\gamma$ is expressed:

$$F(\xi, \eta; \zeta) = \int \frac{e^{-2\pi i\gamma\zeta}}{\gamma}\, e^{-2\pi i(\alpha\xi + \beta\eta)}\, d\alpha\, d\beta$$

With the ancillary plane polars:

$$\alpha = u \cos \psi \qquad \xi = v \cos \varphi$$
$$\beta = u \sin \psi \qquad \eta = v \sin \varphi$$

$$F = \int \frac{e^{-2\pi i\zeta\sqrt{1-u^2}}}{\sqrt{1-u^2}} e^{-2\pi iuv\cos(\psi-\varphi)} u \, du \, d\psi$$

this reduces to:

$$F = 2\pi \int_0^\infty \frac{e^{-2\pi i\zeta\sqrt{1-u^2}}}{\sqrt{1-u^2}} J_0(2\pi uv) u \, du \tag{61}$$

taking into account the standard formula

$$\int_0^{2\pi} e^{iz\cos\psi} \, d\psi = 2\pi J_0(z)$$

Integral (61) was evaluated by Sommerfeld (1909) (see [37] page 416, or [3] page 72). Taking integration trajectory precautions, there follows:

$$\mathscr{F}\left\{\frac{e^{-2\pi i\gamma\zeta}}{\gamma}\right\} = i\frac{e^{-2\pi i\sqrt{\zeta^2+v^2}}}{\sqrt{\zeta^2+v^2}} = i\frac{e^{-2\pi i\rho}}{\rho}$$

By differentiating with respect to the parameter ζ, the following result is found:

$$\mathscr{F}\{e^{-2\pi i\gamma\zeta}\} = -\frac{1}{2\pi}\left(\frac{e^{-2\pi i\rho}}{\rho}\right)_\zeta \tag{62}$$

Returning to (60):

$$\vec{E} = -\frac{1}{2\pi}\vec{E}^{(0)} * \partial_\zeta\left(\frac{e^{-2\pi i\rho}}{\rho}\right) \tag{63}$$

or explicitly

$$\vec{E}(\xi, \eta, \zeta) = -\frac{1}{2\pi}\int \vec{E}^{(0)}(\xi', \eta')\, \partial_\zeta\left(\frac{e^{-2\pi i\rho}}{\rho}\right) d\xi' \, d\eta' \tag{64}$$

where in the integral:

$$\rho = [(\xi - \xi')^2 + (\eta - \eta')^2 + \zeta^2]^{1/2}$$

It is possible to return to the non-reduced variables by noting that ∂_z designates the normal derivative ∂_n:

$$\vec{E}(x, y, z) = -\frac{1}{2\pi}\int \vec{E}^{(0)}(x', y')\, \partial_n\left(\frac{e^{-ikr}}{r}\right) dx' \, dy' \tag{65}$$

(g) Theoretical addendum

Consider, once again, the interpretation of (29) as a Fourier Transform. More generally, the superposition of plane waves, defined by a surface measure $\vec{P}$

on S:

$$\vec{E}(\vec{\rho}) = \int_s e^{-2\pi i\vec{\rho}\cdot\vec{u}}\, d\vec{P}$$
$$\vec{H}(\vec{\rho}) = \frac{1}{\eta}\int_s e^{-2\pi i\vec{\rho}\cdot\vec{u}}\,\vec{u}\wedge d\vec{P} \tag{66}$$

with

$$d\vec{P}(\vec{u})\cdot\vec{u} = 0 \quad (\text{where } \eta = \sqrt{\mu/\varepsilon})$$

The field is the Fourier Transform of a limited measure:

$$\vec{E} = \mathscr{F}\{\vec{P}\} = \langle \vec{P}, e^{-2\pi i\vec{\rho}\cdot\vec{u}}\rangle$$

where

$$\vec{u} = (\alpha, \beta, \gamma) \in \mathbf{R}^3$$

A technique for inversion by continuity is as follows. Let $\theta_n(\vec{u}) = \theta_n(\alpha, \beta, \gamma)$, a regularizing series (approximate unity) $\theta_n \in \mathscr{S}$, $\theta_n \to \delta$ (see Appendix 3), such that: $\vec{P} * \theta_n \to \vec{P} \Rightarrow \mathscr{F}\{\vec{P}\}\mathscr{F}\{\theta_n\} \to \vec{E}$, the convergences being understood in the sense of the tempered distributions.

A series of inverse Fourier Transforms converging towards $\vec{P}$ is formed:

$$\mathscr{F}^{-1}\{\vec{E}\hat{\theta}_n\} = \int \vec{E}(\vec{\rho})\hat{\theta}_n(\vec{\rho})\, e^{2\pi i\vec{\rho}\cdot\vec{u}}\, d\rho \tag{67}$$

where the $\hat{\theta}_n \to 1$ make the integral converge. An example of approximate unity is seen in the Gaussian functions:

$$\theta_n(\vec{u}) = n^3\, e^{-\pi n^2|\vec{u}|^2}, \quad \hat{\theta}_n(\vec{\rho}) = e^{-(\rho^2/n^2)} \tag{68}$$

Generally speaking, it is sufficient to start with an integrable function $\theta(\vec{u})$ whose integral is 1 and whose Fourier Transform is integrable. The approximate unity is: $\theta_n(\vec{u}) = \theta(n\vec{u})$.

B: Approximation of geometrical optics

1: Locally plane waves

(a) Sommerfeld–Runge approximate solutions

The idea of using approximation with locally plane waves (page 50) is a very old one, at least in the scalar case (Sommerfeld and Runge, 1911). Apart from the simplification of calculation, it has the advantage of giving a readily visualized image of propagation, even in a non-homogeneous medium.

In this case, and at very high frequency, it can be assumed that the characteristics of the medium, ε, μ, n, η, vary slowly over paths of the order $\lambda_0 = 2\pi/k_0$, the wavelength *in vacuo* ($\lambda(x)$ can no longer be taken as reference as it varies with the medium). A 'small' domain around point x is therefore, generally, practically homogeneous. It is reasonable to consider pieces of plane waves there. By connecting these along the surfaces of the waves, an easily visualized image of the propagation can be given.

This is why calculations are based on the hypothesis that a single phase $\omega t - \varphi(x)$ of the fields is propagated, carried by the wave surfaces (equiphases) $\varphi(x) = \text{constant}$. The variations in the phase are rapid, and for this reason it is convenient to state:

$$\Phi(x) = k_0 S(x) \tag{69}$$

where $S(x)$ is the iconal function ($\varepsilon\grave{\iota}\kappa\acute{\omega}\nu$ = image) used by optical scientists.

The approximation for very high frequencies therefore consists of finding asymptotic solutions $k_0 \to \infty$, in the form:

$$\vec{E}(x) = \vec{E}_0(x)\, e^{-ik_0 S(x)} \tag{70}$$

$$\vec{H}(x) = \vec{H}_0(x)\, e^{-ik_0 S(x)} \tag{71}$$

of Maxwell's equations, E_0, H_0 and S being slowly-varying.

This leads to the phenomena being expressed in the following way:

- Local structure of field.
- Propagation of the wave surfaces $S(x) = \text{constant}$, the description of which, using rays, leads to the results of geometrical optics.
- Transport of the field along the rays: transport of electromagnetic energy, then that of polarization.

(b) Local field structure

Only media with no losses that are linear, isotropic but not homogeneous will be considered here for example: $D(x) = \varepsilon(x)E(x)$, $B(x) = \mu(x)H(x)$.

Outside the sources, by introducing: $\vec{C}_0(x) \exp[-ik_0 S(x)]$ into (91) and (92) of Chapter 1, the following is found:

$$\text{curl } \vec{C}_0 - ik_0 \text{ grad } S \wedge \vec{C}_0 = ijk\vec{C}_0 \tag{72}$$

$$\text{div } \vec{C}_0 - ik_0 \text{ grad } S \cdot \vec{C}_0 = 0 \tag{73}$$

By introducing the refractive index of the medium $n(x) = k/k_0$ and the unit vector $\vec{\tau}$ such that:

$$\text{grad } S = |\text{grad } S|\vec{\tau},$$

these equations can be reduced, for k_0 and $k \to \infty$, to:

$$|\text{grad } S|\vec{r} \wedge \vec{C}_0 = -jn\vec{C}_0 \tag{74}$$

$$\vec{\tau} \cdot \vec{C}_0 = 0 \tag{75}$$

from which it is deduced that:

$$|\text{grad } S| = n \tag{76}$$

and that the fields E_0 and H_0 are orthogonal to each other and to the direction of propagation. Locally, the structure is that of a homogeneous plane wave;

hence, the local Poynting vector:

$$\vec{P} = \frac{1}{2}\vec{E} \wedge \vec{H}^* = \frac{1}{2\eta}(\vec{E}_0 \cdot \vec{E}_0^*) = \frac{E'^2}{2\eta}\vec{\tau} \tag{77}$$

Notes

• It should be noted that the method is not valid in areas where E_0 and H_0 vary too rapidly: sudden changes of medium, 'shadows' of opaque bodies, caustics, etc.

• More generally, expansions can be considered in the form:

$$\vec{E}(x, k_0) = e^{-ik_0 S(x)} \sum_{n=0}^{\infty} \frac{\vec{E}_n(x)}{k_0^n}$$

and the field can be described by recurrence of successive terms in the series ([24], [19]). The Sommerfeld–Runge approximation is that of the first term. It is known as the geometrical optics approximation, although its content is far more extensive (polarization and phase).

• The far-field radiation of antennas $k_0 r \gg 1$, is also that for which $k_0 \to \infty$. It is therefore not surprising that the local plane wave structure is also present.

(c) Wave surfaces

These are determined by the iconal equation (76), with partial derivatives $(\partial_i S)$ of the first order. Thus, vectorially:

$$\text{grad } S = n\vec{\tau} \tag{78}$$

The wave surfaces are surfaces S = constant of the solutions to this equation. $\vec{\tau}$ is a unit normal on such a surface. Without entering into too much discussion, the similarity, already recognized in the 19th century, that it reveals between analytical mechanics and geometrical optics (Hamilton–Jacobi) must be mentioned. (See, for example Luneburg [24], Born and Wolf [6]).

Very briefly, the following can be observed. Let $S = S_0$ (real), a wave surface. The points on a very close wave surface $S = S_0 + h (h > 0)$ are situated, according to (78), on segment of the normals to the first, with lengths h/n. This leads to a step by step (Huygens) construction, giving a model of propagation and revealing the role of the trajectories orthogonal to the family of wave surfaces thus obtained.

In a homogeneous medium, n = constant, propagation is by parallel wave surfaces.

2: Rays

(a) Ray equation

Consider a family (with one parameter) of wave surfaces. The family (with two parameters) of their orthogonal trajectories is a family of rays. In an isotropic medium, the rays are the field lines of Poynting vectors. They have an important physical significance, in that they 'illustrate' the transport of electromagnetic energy. It is easy to obtain the first-order differential equation that governs them:

$$\frac{d}{ds}(n\vec{r}) = \text{grad } n \tag{79}$$

on each ray parametrized by its arc s.

Note

A fictitious time $dt = ds/n$ allows each ray to be considered as the trajectory of a particle of unit mass, with velocity grad S and acceleration grad $(n^2/2)$, and therefore in a force field deriving from the potential $-n^2/2$. This accounts for the immediate properties of the rays.

(b) Properties of rays

(1) Curvature If ν designates the unit principal normal, and ρ the radius of curvature, with normal acceleration v^2/ρ:

$$\frac{1}{\rho} = \frac{\text{grad } n}{n} \cdot \vec{\nu} = \partial_n \log n \tag{80}$$

Grad n is situated in each osculator plane on the concave side (rays are said to curve towards increasing indices).

The curvature is zero in the very important case of homogeneous media. The rays are then rectilinear.

(2) Media with spherical symmetry The index $n(r)$ depends only on the distance r from a fixed origin (for example, the ionosphere, spherical lenses, etc). In the mechanical analogue, motion is centrally accelerated. It is plane and satisfies the area characteristic, $r^2 d\theta/dt$ = constant, in plane polar coordinates r, θ. Taking into account the velocity value $ds/dt = n$, Bouguer's law can be deduced:

$$rn \sin \alpha = \text{constant} \tag{81}$$

where α is the angle of the tangent to a trajectory with the vector ray.

Exercises

- Deduce from (81) the differential equation (with separate variables) of the rays, giving $\theta(r)$.
- Do the same for the case of layered media: $n = n(z)$. Discuss the various types of ray.

(3) Fermat's principle With the fictitious kinetic energy $T = mv^2/2 = n^2/2$ and the potential $V = -n^2/2$, the Lagrange function $L = T - V = n^2$, is formed.

The principle of stationary action, $\delta \int_{t_1}^{t_2} L\,dt = 0$, is transformed into Fermat's stationary optical path

$$\delta \int_{s_1}^{s_2} n\,ds = 0 \tag{82}$$

which is interesting for a number of reasons: (a) it can be taken as an axiom of geometric optics; (b) it gives the rays the significance of geodesics in the Riemannian space obtained by attaching the 'length' $d\sigma = n|dx|$ to each short vector $(x, x + dx)$; (c) an intuitive extension is used as the basis for the geometric theory of diffraction.

(c) Applications to antennas

The first applications are simply a transposition of standard geometric optics to microwaves with, however, a great deal of variety in the propagation media.

• Parabolic reflector: The point at infinity on the axis of a paraboloid of revolution is the stigmatic image of the focus. Reflectors are the most widely used components of antennas at microwave frequencies. A large part of this work is dedicated to them.

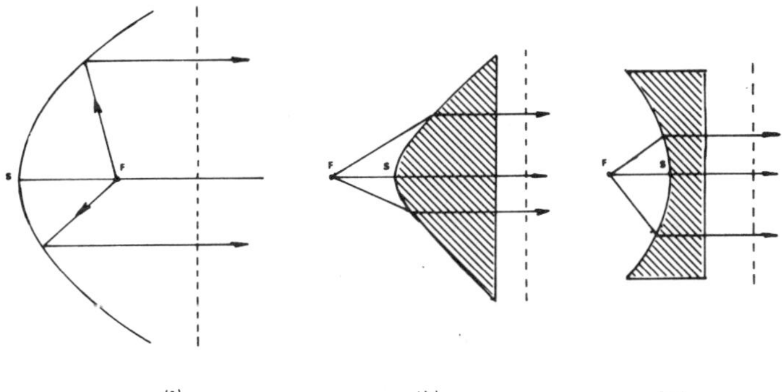

Fig. 8. *Mirrors and lenses.*

• Lenses: It can be verified immediately that for any planoconvex or planoconcave structure, whose curved surface is formed by a quadric of revolution of focus at the origin and eccentricity n (Figure 8), the focus is imaged, stigmatically, at infinity.

The case of $n > 1$ (hyperboloid) is that of dielectrics:

$$n = \sqrt{\varepsilon_r} \tag{83}$$

The case of $n < 1$ (ellipsoid) is encountered in artificial dielectrics. The simplest example is that of a stack of conductive plane sheets regularly spaced at a, guiding a wave of type H_1.

This results in:

$$n = \sqrt{1 - \frac{\lambda^2}{4a^2}} \tag{84}$$

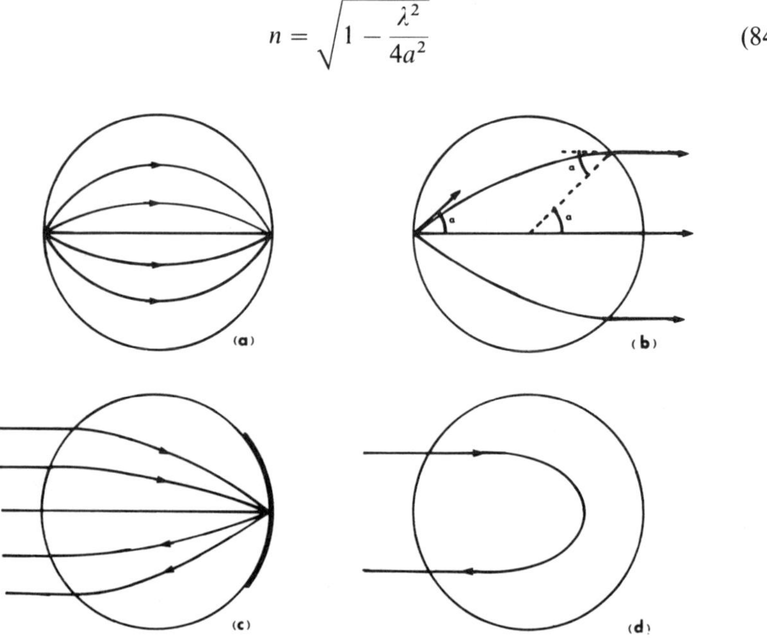

Fig. 9. *Spherical lenses.*

• Spherical lenses: The first example (fish-eye) was given by Maxwell with the index characteristic:

$$n(r) = \frac{2}{1 + \dfrac{r^2}{R^2}} \tag{85}$$

The rays are governed by the equation in Exercise 1. Each point on the surface of the sphere $r = R$ is the stigmatic image of the point diametrically opposite. The rays are circular arcs (Figure 9a).

Luneburg ([24]) reintroduced this subject 30 years ago. The progress made in dielectrics (polythene, Teflon) and the ease of production (for example, in layers), together with the use of artificial dielectrics, made these lenses popular.

The medium

$$n(r) = \sqrt{2 - \frac{r^2}{R^2}} \tag{86}$$

provides a lens that forms an image of any point on the surface at infinity, stigmatically (Figure 9b), hence the possibility of full and rapid scanning without deformation.

Exercise
Verify that the rays are elliptical arcs.

Note
With part of the surface made to reflect, a catadioptric obstacle is formed (Figure 9c). This is avoided in Eaton's lens (Figure 9d):

$$n(r) = \sqrt{\frac{2R}{r} - 1} \tag{87}$$

These passive structures have obvious applications: radar tests, location of space vehicles, etc.

• Guided-ware lenses: Another method consists of employing propagation between two plane parallel conductors, to which an appropriate inward curvature is given. The rays are geodesics of the median surface. Finding them is a standard problem in differential geometry.

This technique is very flexible, and is well suited to the conversion of curvilinear source, by marrying up the appropriate contours on the edges of the entrance and exit of the parallel plate guide.

With a surface, one edge of which is circular, the other rectilinear, it is possible to convert the excitation from a primary source which revolves continuously in front of the curved surface into alternating scanning from the second. A great many such systems were designed during the last war (Robinson).

One variant consists of using a guide made up of two conductors not parallel to each other. As in (84), use is made of the variable gap a to simulate a medium of prescribed index law.

3: Field transport

(a) Transport equation

The transport of the e.m. field is its propagation along the rays. The phase propagation is derived immediately. Along the ray reference arc s, (89) and (98) apply, such that:

$$\varphi(s) - \varphi(s_0) = \int_{s_0}^{s} k_0 n(s')\, ds' = \int_{s_0}^{s} k(s')\, ds' \tag{88}$$

In a homogeneous medium $\varphi(s) - \varphi(s_0) = k(s - s_0)$ on the rays which are then rectilinear. It remains to study the transport of fields $\vec{E}_0$ and $\vec{H}_0$, which are best considered in the form

$$\vec{E}_0 = E'_0\, \vec{e}_0, \quad \vec{H}_0 = H'_0 \vec{h}_0 \tag{89}$$

This separates the norms E'_0 and H'_0 (Chapter 1, (50)), from the normalized fields $\vec{e}_0$ and $\vec{h}_0$. This comes down to distinguishing between the transport of a power density (77) and that of a polarization.

Starting with equations (72) and (73), the calculations can be expressed without difficulty, although they require a good understanding of straightforward vector analysis (see [6] or [24]). They result in a transport equation, which a simple calculation separates into a transport equation for Poynting's vector (77):

$$\frac{d}{ds}\log P = 2\Gamma \tag{90}$$

where Γ designates, at each point, the mean curvature

$$\Gamma = \frac{1}{2}\left(\frac{1}{R_1} + \frac{1}{R_2}\right) \tag{91}$$

of the wave surface that passes through this point (the radii of principal curvature R_1 and R_2 are considered algebraically relative to τ) and a polarization transport equation:

$$\frac{d\vec{e}_0}{ds} = -(\vec{e}_0 \cdot \text{grad} \log n)\vec{\tau} \tag{92}$$

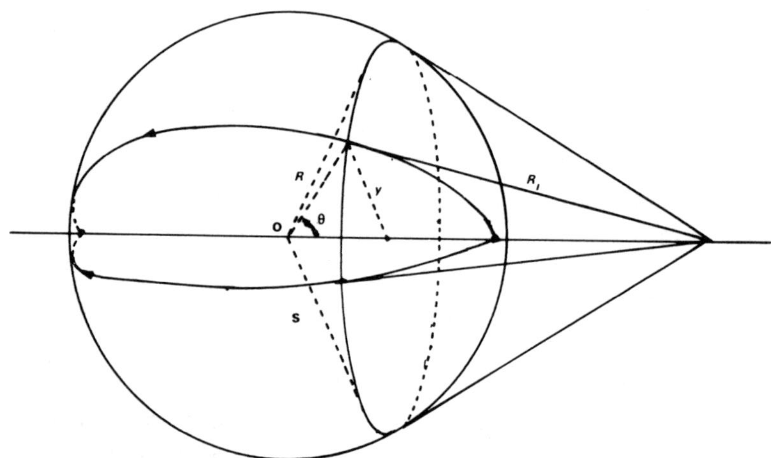

Fig. 10. *Propagation on a sphere.*

(b) Power transport
Along a ray:

$$P(s) = P(s_0)\exp\left(2\int_{s_0}^{s} \Gamma(s')\, ds'\right) \tag{93}$$

or

$$\frac{E_0'(s)\exp\left(-\int \Gamma\, ds'\right)}{\sqrt{\eta(s)}} = \text{constant} \tag{94}$$

Exercise
Propagation along the meridians of a sphere $r = R$ (Figure 10). Show that $E'_0(s)\sqrt{y/\eta}$ = constant (using Meusnier's theorem).

Note
The conservation of power along a simple tube of rays, with section $\sigma(s)$, is expressed by $P(s)\sigma(s)$ = constant, or:

$$\sigma(s)\exp\left(2\int \Gamma\, ds'\right) = \text{constant} \tag{95}$$

(c) Polarization transport
This is of less practical importance. Only e_0 and h_0 with rectilinear polarization will be considered here. It can be deduced (92) that the transport, along a ray, of the polarization trihedral frame τ, e_0, h_0, causes it to rotate about the tangent τ, so as to correct the rotation of the osculation plane. In the language of Riemann geometry, this is 'parallel transport', in the sense understood by Levi-Civita and Cartan, along the geodesic which is the ray (see page 63). The polarization trihedral frame τ, e_0, h_0 is transported parallel to the rays.

Examples

• Plane rays: the polarization trihedral frame remains fixed relative to Serret's trihedral plane of the ray.

• Rectilinear rays: This concerns a normal euclidian space without curvature, and the transport is an ordinary translation.

4: Homogeneous media

(a) Field transport
The rays are rectilinear. Let S be a wave surface. Each of its normals is a ray. One of them is parametrized by the arc s with $s = 0$ on S. The two principal centres of curvature of S are, also, on the ray considered, those of surfaces parallel to S.

With the convention of measuring the principal radii of curvature, R_1 and R_2, positively towards decreasing s, $R_1 + s$ and $R_2 + s$ are those of the wave surface parallel to S passing through point s.

Thus, the algebraic mean curvature:

$$\Gamma(s) = -\frac{1}{2}\left(\frac{1}{R_1 + s} + \frac{1}{R_2 + s}\right) \tag{96}$$

and its primitive

$$\int \Gamma(s')\, ds' = \frac{1}{2}\log|K(s)|$$

where

$$K(s) = \frac{1}{(R_1 + s)(R_2 + s)} \tag{97}$$

is the total curvature, algebraically. (94) can be reduced to:

$$E'_0(s) = \text{constant} \cdot \sqrt{|K(s)|} \tag{98}$$

or more explicitly:

$$E'_0(s) = E'_0(0)\sqrt{\left|\frac{R_1 R_2}{(R_1 + s)(R_2 + s)}\right|} \tag{99}$$

The parallel transport of the complex electric field along the ray is expressed by:

$$\vec{E}(s) = \vec{E}(0)\sqrt{\frac{R_1 R_2}{(R_1 + s)(R_2 + s)}}\; e^{-iks} \tag{100}$$

The direct use of total curvature and not of their absolute values is not justified by the above, but will be at a later stage (approximation of physical optics).

At a great distance, $s > 0$:

$$\vec{E}(s) = \vec{E}(0)\sqrt{R_1 R_2}\,\frac{e^{-iks}}{s} + o\left(\frac{1}{s}\right) \tag{101}$$

Notes

- For a pencil of rays the following can be deduced from (95): $\sigma(s)/|(R_1 + s)(R_2 + s)| = \text{constant}$, the evident geometric significance of which would allow (99) to be established directly.
- The field (100) would be infinite at the principal centres of curvature (focal points), $s = -R_1, -R_2$. It has already been pointed out that the geometrical optics approach is not valid near the caustic (see below), which separates an area occupied by the rays from an area in shadow. Another approach is necessary, using physical optics.
- Nevertheless, it is often convenient to take one of the foci, F_1 for example, as the origin. $R_1 = 0$ and $R_2 = \rho$, distances from the foci, where:

$$\vec{E}(s)\sqrt{s(s + \rho)}\, e^{iks} = \text{constant}$$

The field is infinite at the origin, varying as $1/\sqrt{s}$. It suffices to write $E(s)\sqrt{s} = F(s)$, such that:

$$\vec{E}(s) = \vec{F}(0)\sqrt{\frac{\rho}{s(s + \rho)}}\, e^{-iks} \tag{102}$$

- For plane problems one of the radii of curvature can be taken. Thus:

$$\vec{E}(s) = \vec{E}(0)\sqrt{\frac{R}{R + s}}\, e^{-iks} = \vec{E}(0)\sqrt{R}\,\frac{e^{-iks}}{\sqrt{s}} + o\left(\frac{1}{s}\right) \tag{103}$$

(b) Caustic and focal lines

Some of the geometric properties of the surfaces should be summarized, particularly as they are hardly ever included in lecture courses.

In a homogeneous domain, the rays are the orthogonal, rectilinear trajectories of families of surface waves. They have remarkable properties of the congruences of normals.

Any wave surface S has two families of orthogonal curves (lines of curvature) having evolutes, the focal curves. This means that the normals to S along the curve line envelop a curve which they each touch at one point, said to be focal. On each normal there are, therefore, two focal points, F_1 and F_2, located by the principal radii of curvature, R_1 and R_2.

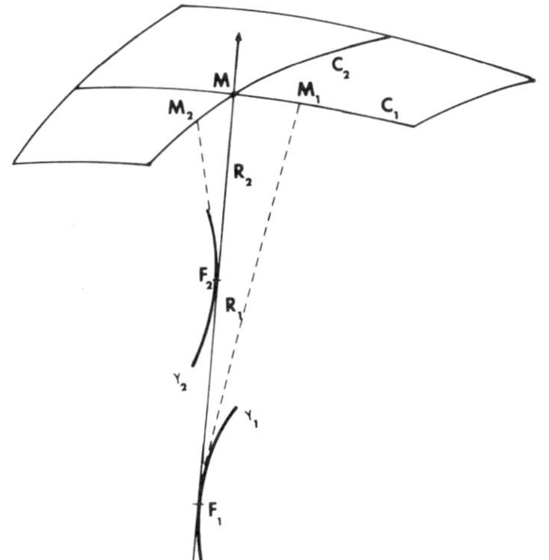

Fig. 11. *Foci and caustic.*

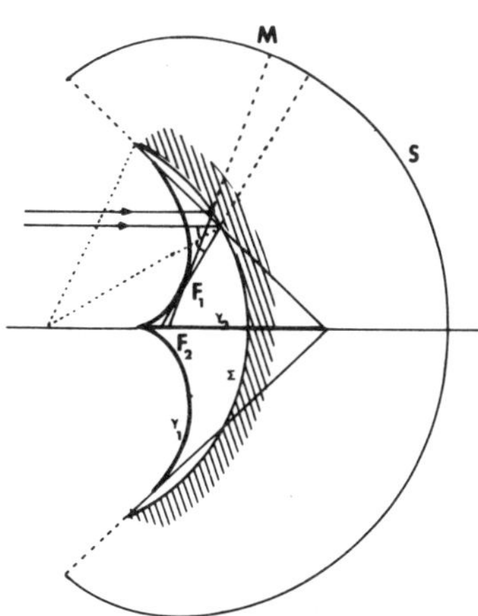

Fig. 12. *Caustic of spherical mirror (plane incident wave).*

The focal curves γ_1 and γ_2 generate the two sheets of a focal surface, the caustic surface of optical scientists. The congruences of the normals have the characteristic property that, through each ray, there are two rectangular planes tangential to the caustic.

One of the sheets of the caustic surface can be reduced to a curve which the rays intersect. This is the case with wave surfaces of revolution, in which the lines of curvature are meridians and parallels. The caustic surface is made up of part of the axis and the surface produced by rotation of the evolute of a meridian.

The two sheets can be two curves (for example, a torus).

The beam of rays is the congruence of straight lines supported by the two curves.

An important specific case concerns homocentric congruences (stigmatism). The rays pass by a single focal point to which the caustic surface is then reduced. The wave surfaces are concentric spheres.

5: Application to reflectors

(a) Reflection on a conductive plane

S is a perfect conductive plane with unit normal $\vec{n}$, and field:

$$\vec{E} = \vec{E}_s + \vec{E}_n \tag{104}$$

resolved into a part parallel to the plane, $\vec{E}_s = (\vec{n} \wedge \vec{E}) \wedge \vec{n}$, and a part normal to it, $\vec{E}_n = (\vec{n} \cdot \vec{E})\vec{n}$, with which is associated the symmetrical transformation:

$$\underline{\vec{E}} = \vec{E}_s - \vec{E}_n \tag{105}$$

For the same conductor, with an incident field $E^{(i)}$, $H^{(i)}$ there is a corresponding, reflected field $E^{(r)}$, $H^{(r)}$, such that the following boundary conditions are found (page 22):

$$\vec{E}^{(r)} = -\underline{\vec{E}}, \quad \vec{H}^{(r)} = \underline{\vec{H}} \tag{106}$$

(b) Geometrical optics of reflectors

Consider a beam of rays incident on the surface Σ of a perfect reflector. One of the rays, parametrized by r, encounters Σ at the point $r = r_0$. Let $K^{(i)}(r)$ be the total curvature of the incident wave surface, transported by the ray. The incident field, according to (100), is:

$$\vec{E}^{(i)}(r) = \vec{E}^{(i)}(0) \sqrt{\frac{K^{(i)}(r)}{K^{(i)}(0)}}\, e^{-ikr} \tag{107}$$

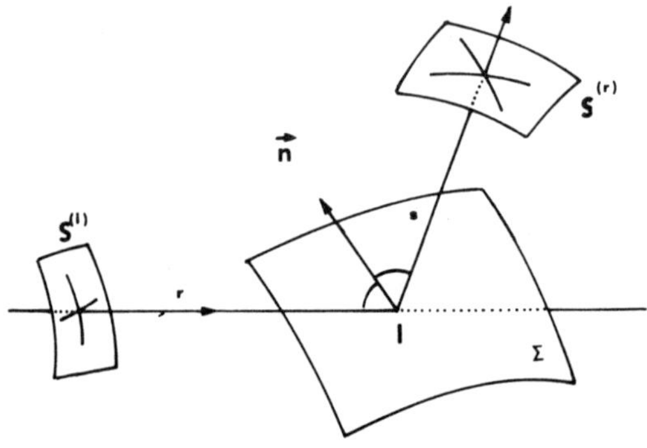

Fig. 13. *Wave surface reflections.*

The reflected ray, parametrized by s, measured from the point of incidence, transports a reflected wave surface with total curvature $K^{(r)}(s)$.

The reflected field is:

$$\vec{E}^{(r)}(s) = \vec{E}^{(r)}(0) \sqrt{\frac{K^{(r)}(s)}{K^{(r)}(0)}}\, e^{-iks} \tag{108}$$

When the point of incidence is regular, the reflection laws are the same as for the plane tangential to the reflector at that point. So, according to (106):

$$\vec{E}^{(r)}(s) = -\vec{E}^{(i)}(0)\sqrt{D\frac{K^{(r)}(s)}{K^{(i)}(0)}}\, e^{-ik(s+r_0)} \tag{109}$$

with the (algebraic) divergence coefficient:

$$D = \frac{K^{(i)}(r_0)}{K^{(r)}(0)} \tag{110}$$

the ratio of total curvatures of the two wave surfaces at the point of incidence. It characterizes the influence of the reflector curvature on the convergence of incident beams. $D = 1$ for a plane reflector. The calculation of D is complicated, even in the particular case of a point primary source at $r = 0$ (see, for example [9], page 498, or [32], page 143).

The divergence coefficient has the following interesting significance. For large values of r (beyond the reflector) and s:

$$\begin{aligned} \vec{E}^{(i)}(r) &\simeq \frac{\vec{E}^{(i)}(0)}{\sqrt{K^{(i)}(0)}}\,\frac{e^{-ikr}}{r} \\ \vec{E}^{(r)}(s) &\simeq -\frac{\vec{E}^{(i)}(0)}{\sqrt{K^{(i)}(0)}}\sqrt{D}\,\frac{e^{-ik(s+r_0)}}{s} \end{aligned} \tag{111}$$

which shows that D is the ratio of radiation characteristics (in the directions defined by reflection) of the antenna, which constitutes the reflector, and of the primary source.

(c) Cylindrical reflectors

The study of cylindrical reflectors with linear primary sources, parallel to the generating lines, is a plane problem that is easy to solve with polar coordinates. With the notation of Figure 14:

$$i = (\pi/2) - V,\ \varphi = (\pi/2) + \theta + V,\ \psi = \theta + 2V$$

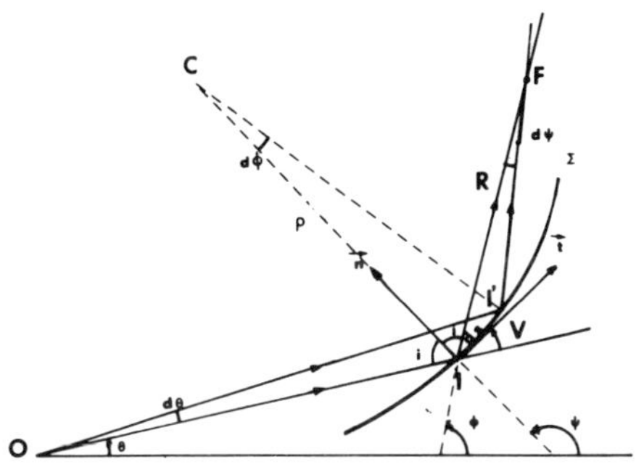

Fig. 14. *Cylindrical mirrors.*

The radius of curvature of the reflector is $\rho = ds/d\varphi = ds/d\theta + dV$ and that of the reflected wave surface is $R = ds \sin V/d\psi = ds \cos i/d\theta + 2\, dV$. A simple calculation gives:

$$D = \frac{R}{r_0} = \frac{\rho \cos i}{2\, r_0 - \rho \cos i} \tag{112}$$

Exercise
Verify that for large values of r and $s > 0$:

$$\vec{E}^{(i)}(r) = \vec{E}^{(i)}(r_0)\sqrt{r_0}\,\frac{e^{-ik(r-r_0)}}{\sqrt{r}} = \vec{F}^{(i)}(\theta)\,\frac{e^{-ikr}}{\sqrt{r}} \tag{113}$$

$$\vec{E}^{(r)}(s) = -\underline{\vec{E}}^{(i)}(r_0)\sqrt{-Dr_0}\,\frac{e^{-iks}}{\sqrt{s}} = \vec{F}^{(r)}(\psi)\,\frac{e^{-iks}}{\sqrt{s}} \tag{114}$$

and hence derive the relationships between the radiation characteristics:

$$|\vec{F}^{(r)}(\psi)|^2 = |D|\,|\vec{F}^{(i)}(\theta)|^2 \tag{115}$$

$$|F^{(r)}(\psi)|^2\,|d\psi| = |\vec{F}^{(i)}(\theta)|^2\, d\theta \tag{116}$$

(conservation of power).

Application: Synthesis of reflectors. Given the primary source $F^{(i)}(\theta)$ and the secondary radiations $F^{(r)}(\psi)$, how can the characteristics of the reflectors be calculated? The differential equation given by (116) can be integrated. Thus:

$$r_0(\theta) = r_0(\theta_0) \exp \int_{\theta_0}^{\theta} \cos \frac{\psi(\theta') - \theta'}{2}\, d\theta' \tag{117}$$

C: Physical optics approximations

1: General points

Physical optics approximations generally start with representations of the Huygens-type e.m. field (the equivalence theorem, page 23 above, Kirchhoff's formulae, page 212, etc); that is, with the values of the fields themselves, on a closed surface surrounding the sources. These representations are rigorous if the surface field, which serves as an intermediary, is known. It may happen, however, that an approximate knowledge of this may be available, though it may break the laws of the e.m. field. This situation arises frequently in calculations concerning antennas. The field at infinity, particularly, is obtained in an acceptable way and with easy calculations. More generally, however, the principal area of application is that of the diffraction of microwaves, which will be discussed in the next chapter.

2: Relationship to geometrical optics

Only homogeneous media will be discussed here. Let S be a closed surface, oriented by its external normal. At each external point x, therefore, Kirchhoff's representation (in bicomplex form) is (Appendix 3 (69)):

$$\vec{C}(x) = \int_S [G(x - x')\, \partial_n \vec{C}(x') - \vec{C}(x')\, \partial_n G(x - x')\, dS \tag{118}$$

where G is the elementary solution (Chapter 1, (136)) and $C(x')$ is the field on the surface.

As on page 59, the field is replaced by the Sommerfeld–Runge approximation, and the wave surface $S(x') = S_0 =$ constant is taken for S. *In vacuo*, the index $n = 1$ and grad S is reduced to the unit normal to this surface (see (76)). Then (118) takes the following, easily obtained form:

$$\vec{C}(x) = -\frac{e^{-i\kappa S_0}}{4\pi} \int_S \vec{F}(x') \frac{e^{-ikr}}{r}\, dS \tag{119}$$

where:

$$\vec{F}(x') = ik(\partial_n r - 1)\vec{C}_0 + o(k) \tag{120}$$

In (119), the vectorial transposition of the Huygens 'wavelets' can be recognized.

The wave surface can be described in ancillary coordinates at one of its points 0, with the normal oriented on the convex side and the principal directions at 0 (Figure 15):

$$w = -\frac{u^2}{2R_1} - \frac{v^2}{2R_2} + o(u^2 + v^2), \quad 0 < R_1 < R_2 \tag{121}$$

At a point s of the normal:

$$r = \sqrt{u^2 + v^2 + (s - w)^2} = \pm\left(s + \frac{R_1 + s}{2R_1 s} u^2 + \frac{R_2 + s}{2R_2 s} v^2 + o(u^2 + v^2)\right)$$

($\pm$ according to the sign of s).

- $s > 0$: at a point situated on the normal, on the convex side, the field (119) can, for $k \to \infty$, be estimated using the stationary phase approximation (Appendix 2, (2)):

$$\vec{C} = \frac{i}{2k} \vec{F}(o) \sqrt{\frac{R_1 R_2}{(R_1 + s)(R_2 + s)}}\, e^{-ik(S_0 + s)} \tag{122}$$

Taking account of the fact that, on S, $\partial_n S = 1$, the following is found:

$$\vec{C}(s) = \vec{C}(o) \sqrt{\frac{R_1 R_2}{(R_1 + s)(R_2 + s)}}\, e^{-iks} \tag{123}$$

agreeing with (100).

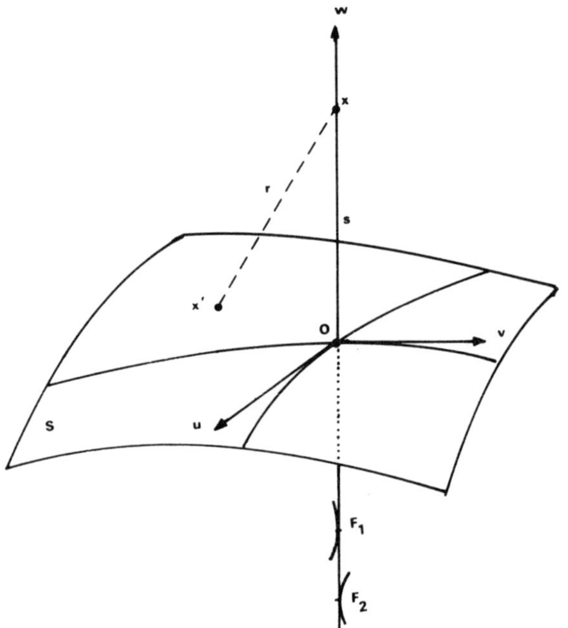

Fig. 15. *Field along a ray.*

This entirely standard calculation clearly indicates that physical optics can be reduced to geometrical optics for $\lambda \to 0$. It is equivalent to dividing S into Fresnel zones about 0. The approximation (122) is the contribution of the central zone, with the effects of the ring zones being destroyed by interference.

- $s < 0$: the result of the previous calculation cannot be extended, without taking precautions, to the concave side of the wave surface. It has no meaning at the focal points $s = -R_1$ or $-R_2$, where, in addition, the expression under the square root changes its sign. By using formula (2) of Appendix 2, the following is obtained:

$$\vec{C}(s) = i^q \vec{C}(o) \sqrt{\frac{R_1 R_2}{|(R_1 + s)(R_2 + s)|}}\, e^{-iks} \tag{124}$$

where $q = 0, 1, 2$ depending on whether $s > -R_1$, $-R_2 < s < -R_1$, $s < -R_2$. The successive phase shifts of $\pi/2$ at the focal points constitutes Gouy's law.

3: The field at the foci

Physical optics allows the behaviour of the field around the caustic surface to be described in detail. Without entering into complicated calculations (for example [9] page 537 or [32] page 447), it is interesting to comment on the results.

The development (121) must be extended to third-order terms, and where the

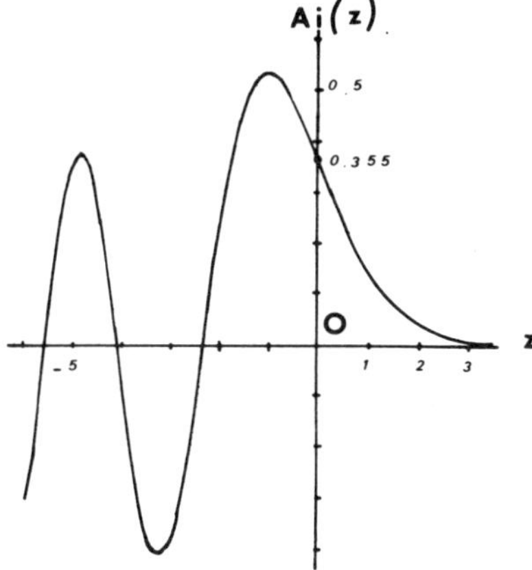

Fig. 16. *Airy's function.*

stationary phase method reduced the calculation of the field to the product of two Fresnel integrals, one of them must be expressed using Airy's function:

$$Aiz = \frac{1}{2\pi}\int_{-\infty}^{\infty} e^{\pm i[(t^3/3)+zt]}\,dt = \frac{1}{\pi}\int_{0}^{\infty}\cos\left(\frac{t^3}{3}+zt\right)dt \qquad (125)$$

The field at focus F_1, for example, is then proportional to $k^{1/6}Ai(0)$. It is finite, but becomes infinite with k, which conforms to the principles of geometrical optics. In addition, around the caustic surface—for example at the point $u = h$, $v = 0$, $w = -R_1$—the field obtained is proportional to $Ai[(2k/\rho_1)^{1/3}h]$ where ρ_1 is the radius of curvature of the caustic F_1. The graph of Airy's function (Figure 16) shows that, contrary to geometrical optics, physical optics allows for penetration of the field into the 'obscured' area ($h > 0$), which is limited by the caustic surface. In addition, the exponential decrease is reminiscent of the behaviour of evanescent waves, and suggests, once again, that the complex rays should be put to good use. The sinusoidal appearance of the amplitudes of the field components should be noted in the 'illuminated' area ($h < 0$) around the caustic.

Chapter 3

Introduction to diffraction theory

A: General points

1: The role of electromagnetic diffraction

The diffraction of light was first observed among phenomena which geometrical optics could not account for (penetration of light into the geometric shadow of obstacles). It was connected more with physical optics, that is, the wave nature of light.

In acoustics and then electromagnetism interest was taken in the general problem of the disturbance to which the propagation of waves is subjected by obstacles of all kinds. From the start of the post-Hertzian period, propagation around the earth came to be regarded as a phenomenon of diffraction.

It is remarkable that just before the First World War a number of so-called canonical problems were solved. The names of Rayleigh, Mie, Debye, Sommerfeld and Poincaré indicate the importance attached to the subject.

In the study of antenna construction, coupled elements, such as reflectors and directors (Yagi antennas) or mechanical supports (shrouds, mountings, etc) can, certainly, be considered as obstacles to radiation. Current numerical techniques allow the further research on the subject of diffraction.

It is however, in relation to microwaves that the role of diffraction is the most remarkable (the radar problem is typical). Antennas in this range are often inspired by optical science, and it is not surprising that present-day techniques make great use of the theory of diffraction in a rigorous treatment of problems concerning reflectors, lenses, radiating apertures, etc. The simplest, although still difficult, case will be considered here: that of perfectly-conducting obstacles in a perfect medium.

A diffraction problem involves the study of the e.m. field $\vec{C}$ created by sources in the presence of obstacles. If the reaction of the obstacles to the sources is ignored, the so-called 'total' field can be treated as the superposition

$$\vec{C} = \vec{C}^{(i)} + \vec{C}^{(d)} \tag{1}$$

of field $\vec{C}^{(i)}$ (the incident field) that the source creates in the absence of any obstacles and of the field $\vec{C}^{(d)}$ (the diffracted field). The incident field is the known part of the problem, the diffracted field is the unknown. The total field, equal to zero inside perfectly conducting obstacles, obeys the equivalent boundary conditions on their boundary S:

$$\vec{n} \wedge \vec{E}_S = 0, \quad \vec{n} \cdot \vec{H}_S = 0 \tag{2}$$

The diffracted field is an *external* solution to Maxwell's equations. The uniqueness of the solution, for a bounded set of obstacles, is ensured by considering the radiation conditions at infinity (page 39) and, possibly, by the finite energy conditions on the ridges and peaks (page 22).

The incident field is often that of a homogeneous plane wave. It does not obey the radiation conditions and consequently the total field does not do so either. The surface charges and currents induced on the positive side of S are given by:

$$\begin{aligned} \rho_S &= \frac{1}{\varepsilon}\vec{n} \cdot \vec{E}_S = \frac{1}{\varepsilon}\vec{n} \cdot (\vec{E}_S^{(i)} + \vec{E}_S^{(d)}) \\ \vec{J}_S &= \vec{n} \wedge \vec{H}_S = \vec{n} \wedge (\vec{H}_S^{(i)} + \vec{H}_S^{(d)}) \end{aligned} \tag{3}$$

and constitute secondary sources, the diffracted field from which can be calculated. Determining these intermediaries is equivalent to the integral equation approach (see below).

2: Solving diffraction problems

(a) Exact solution

Solving diffraction problems involves initially the solution of partial differential equations with boundary conditions, the difficulty of which—compared, for example, with those of acoustics—is due to the vector nature of the magnitudes involved.

There are very few obstacles that lend themselves to exact solution in the sense of standard analysis, which usually means a correct use of special functions. They have been known for a long time, and it may be tempting to see them only as the object of academic exercises. They provide, however, reference results which allow the methods applied to be proved, or semi-empirical techniques, such as the geometrical theory of diffraction (GTD), to be clarified (see below). They give a precise idea of local behaviour: singularities on ridges, creeping around the convex boundaries, etc. Finally, the advantage that can be derived from cylindrical cases, which are treated in a scalar, two-dimensional way, will be retained. These are equivalent to acoustic problems, and the numerical processing involved is far simpler. Some notes are given below on the example of the cylinder of revolution and the half-plane.

(b) Approximate solutions

The problem with 'exact' solutions is common to all mathematical physics, and many methods of approximation have been developed (see, for example [15], [20], [22]). One example is the perturbation of boundaries, a famous illustration of which can be seen in the treatment of the cylindrical antenna by the very thin prolate spheroid (Max Abraham, 1901; Stratton and Chu, 1942). Above all, mention must be made of the asymptotic expansions to which sinusoidal fields lend themselves so well when the wavelength is large or small relative to the dimensions of the obstacle. The second case, which will be considered later, often arises in microwaves. Use is made of the representations of physical optics (Kirchhoff's approximation) and geometrical optics, as explained in Chapter 2. The geometrical theory of diffraction is basically a consideration of diffracted rays.

(c) Numerical solutions

The computer has totally altered ideas accepted until recently concerning the numerical solution of engineering problems and has even opened new avenues of research that are now very active.

The most general problems, of arbitrary incident waves, and obstacles of any form with dimensions of the order λ (sometimes known as the resonance case), have become accessible.

In parallel with analysis, one has a dual approach: differential or integral. The role of the computer in problem-solving using integral equations, which corresponds to a very natural formulation of diffraction problems, has been decisive. These equations are now familiar to all engineers.

Finally, extremely difficult problems, such as that of diffraction by n obstacles, have been reconsidered in an original and effective way.

The limitations of numerical calculations must not, however, be underestimated: limited capacities make three-dimensional problems difficult, and there may be errors arising from rounding of numbers, convergences, and cases of non-uniqueness, etc.

B: Two canonical examples

1: Cylinder of revolution (Rayleigh, 1881)

The semipolar coordinates (r, θ, z) are used. The conductor is the cylinder $r = a$ and the incident wave is a homogeneous plane wave with propagation vector (k, π, o). The two correlative cases of the electric and magnetic fields, polarized parallel to the axis, can both be treated. Only the former will be considered here:

$$E_z^{(i)} = e^{ikr\cos\theta} = \sum_{n=-\infty}^{\infty} i^n J_n(kr)\, e^{in\theta} \tag{4}$$

using a standard expansion.

The diffracted field is expressed by the series of particular solutions to Helmholtz's equation, obeying the radiation conditions (for $\mathbf{R}^2$):

$$E_z^{(d)} = \sum_{n=-\infty}^{\infty} A_n H_n^{(2)}(kr)\, e^{in\theta} \tag{5}$$

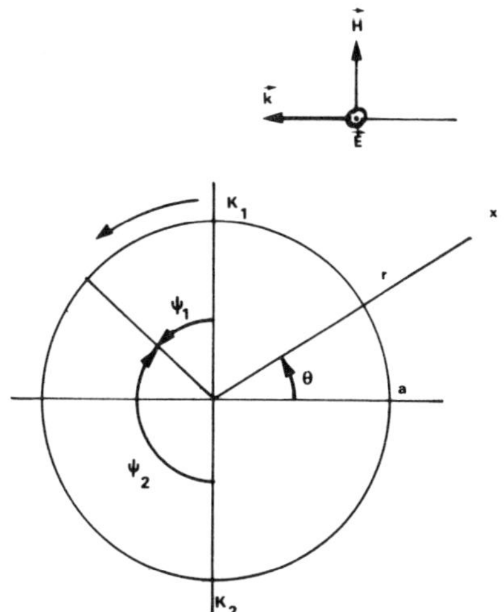

Fig. 17. *Diffraction by a cylinder of revolution.*

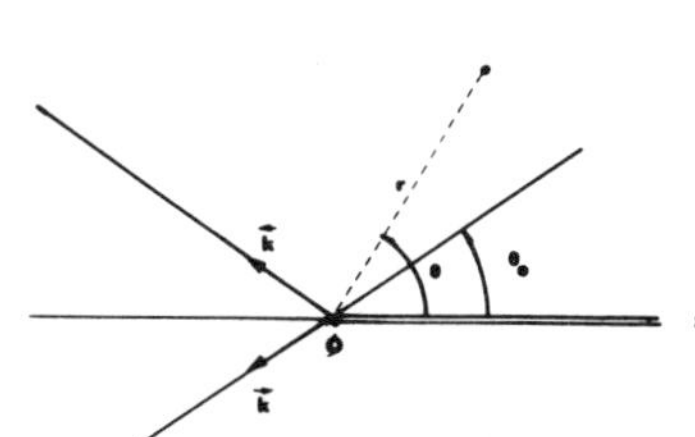

Fig. 18. *Diffraction by a half-plane conductor.*

The boundary condition $E_z^{(i)} + E_z^{(d)} = 0$ for $r = a$ provides the coefficients:

$$A_n = -i^n \frac{J_n(ka)}{H_n^{(2)}(ka)} \tag{6}$$

At infinity, notably, the following will be found:

$$E_z \simeq -\sqrt{\frac{2}{\pi kr}}\, e^{-i[kr-(\pi/4)]} \sum_n (-1)^n \frac{J_n(ka)}{H_n^{(2)}(ka)}\, e^{in\theta} \tag{7}$$

The magnetic field, calculated with the first Maxwell equation, gives the surface current:

$$J_z = H_\theta^{(i)} + H_\theta^{(d)} = \frac{2}{\eta\pi ka} \sum_n \frac{i^n}{H_n^{(2)}(ka)}\, e^{in\theta} \tag{8}$$

This is the simplest example of a diffraction calculation. Unfortunately, the series converges less rapidly when ka is large. The order of magnitude $J_n(n) \sim 0.45/\sqrt[3]{n}$ is significant. The Watson Transform is a frequently used procedure, which consists of treating (8) as a series of residuals.

Exercise

Verify that (8) is the integral of the function of complex v: $-i\exp[iv(\theta - \pi/2)]/\pi ka H_v^{(2)}(ka)\sin \pi v$ on a contour surrounding the poles $v = n$, $n \lessgtr 0$.

Creeping waves

The contribution of geometrical optics and a pure diffraction term that exists only in the 'obscured' part $\pi/2 < \theta < 2\pi/2$ can be revealed in the current. A complicated calculation (see [29], page 120) is used to deduce from the integral mentioned in the exercise the approximate expression of this term, in the form:

$$J_z \simeq \frac{1}{\eta} e^{-i(\pi/6)} \sqrt[3]{\frac{2}{ka}} \sum_{n=-\infty}^{\infty} \sum_{p=0}^{\infty} D_n [e^{-i\nu_n(\psi_1 + 2\pi p)} + e^{-i\nu_n(\psi_2 + 2\pi p)}]$$

with the poles ν_n of $H_\nu^{(2)}(ka)$ considered as an analytical function of the index and the auxiliary angles $\psi_1 = \theta - (\pi/2)$, $\psi_2 = -\theta + (3\pi/2)$ measured positively towards the shadow from its respective boundaries. The contents of the square brackets can be interpreted as the superposition of current waves (known as creeping waves) propagating indefinitely in opposite directions, round the surface of the cylinder and which are also found, in a similar way, in the expression for the field. Complex ν_n brings about a stronger attenuation when ka is large. In practice, only the very first modes survive on one part of their paths. The calculation of the poles ν_n and the coefficients D_n is very difficult. To clarify these ideas, note that the asymptotic expressions $(ka) \gg 1)$ can be derived:

$$\nu_n \simeq ka + e^{i(2\pi/3)} \sqrt[3]{\frac{ka}{2}}\, q_n$$
$$D_n \simeq -e^{-i(\pi/6)} \sqrt[3]{\frac{2}{ka}} \frac{1}{A_i'(q_n)} \tag{9}$$

where the q_n are zeros of Airy's function (see Chapter 2, (125)).

The concept of the creeping wave is far from being trivial. It is an ancillary geometric representation that is of use in GTD, and this has contributed to its attaining a level of concrete physical significance which it otherwise does not merit.

2: Conductive half-plane (Sommerfeld, 1896)

The solution, by Sommerfeld, has given the problem of plane wave diffraction at the edge of a sheet a historical and theoretical importance. But above all, the solution is a basic reference (canonical problem) that can be used in approximate solutions to problems with any edges (Braunbecks's method, GTD, etc).

(a) Polarization E (electric field parallel to edge)

Consider a conductive sheet of negligible thickness, constituting the half-plane $y = 0$, $x > 0$ in cartesian coordinates (x, y, z), or $\theta = 0$, 2π in semipolar coordinates (r, θ, z), $0 \leqslant \theta \leqslant 2\pi$.

The normally incident plane wave $E_z^{(i)} = \exp[ikr\cos(\theta - \theta_0)]$ arrives from direction θ_0. By symmetry, the limitations $0 \leqslant \theta_0 \leqslant \pi$ can be set. Geometrical optics determines, in direction $\pi - \theta_0$, the reflected plane wave $E_z^{(r)} = -\exp[ikr\cos(\theta + \theta_0)]$.

The total field, including the diffraction, can be calculated in a variety of ways, leading to the expression

$$E_z = E_z^{(i)} G(\tau_1) + E_z^{(r)} G(\tau_2) \tag{10}$$

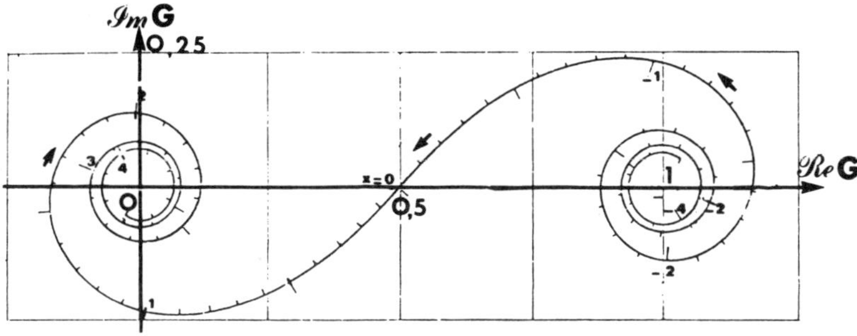

Fig. 19. *G(x) function (Cornu's spiral).*

where

$$\tau_{1,2} = -\sqrt{2kr}\left(\cos\frac{\theta \mp \theta_0}{2}\right)$$

and

$$G(\tau) = \frac{e^{i(\pi/4)}}{\sqrt{\pi}}\int_\tau^\infty e^{-it^2}\,dt = \frac{1}{2} - \frac{e^{i(\pi/4)}}{\sqrt{2}}F^*\left(\sqrt{\frac{2}{\pi}}\,\tau\right) \tag{11}$$

F being the usual Fresnel integral (Appendix 2).

It can be proved that E_z is really zero on the two sides $\theta = 0$ and 2π of the sheet, and it will be noted that the field only reproduces itself periodically, every two revolutions. Each revolution, however, changes its sign. The situation is that of a non-uniform function in the complex plane $x + iy$. Sommerfeld imagines that each plane z = constant is a Riemann surface with two leaves (see above), as though the incident rays passed from one leaf to the other in crossing the conductor.

Exercise

Deduce from (10) the magnetic field, then the current on the sheet. Prove that in the vicinity of the edge

$$J_z = \frac{2}{\eta}\sqrt{\frac{2}{\pi kr}}\,e^{-i(\pi/4)}\sin\frac{\theta_0}{2} + 0(\sqrt{r}) \tag{12}$$

The behaviour of the fields around the edge: verify Meixner's condition (page 22, above).

(b) Far field

The asymptotic expression, $|\tau| \to \infty$,

$$G(\tau) = u(-\tau) + \frac{1}{2\sqrt{\pi}} \frac{e^{-i[r^2+(\pi/4)]}}{\tau} + 0\left(\frac{1}{\tau^3}\right) \tag{13}$$

(u: Heaviside step function) means that for $kr \gg 1$, the following can be established:

$$E_z = u(-\tau_1)E_z^{(i)} + u(-\tau_2)E_z^{(r)} + E_z^{(d)} \tag{14}$$

where

$$E_z^{(d)} = \frac{e^{-i(\pi/4)}}{2\sqrt{2\pi k}}\left(\frac{1}{\cos\dfrac{\theta+\theta_0}{2}} - \frac{1}{\cos\dfrac{\theta-\theta_0}{2}}\right)\frac{e^{-ikr}}{\sqrt{r}} = D_E \frac{e^{-ikr}}{\sqrt{r}} \tag{15}$$

the interpretation of which is clear. The total field, at great distances, is the superposition of the incident field (in the 'illuminated' sector, $0 \leqslant \theta < \pi + \theta_0$), the reflected field (in the sector $0 \leqslant \theta < \pi - \theta_0$) and a purely diffracted field (15) (everywhere). The latter does not correspond to the strict definition (1), but is the only one to obey the radiation condition. It can be interpreted as a cylindrical wave which would be 'radiated' at infinity by the ridge with a radiation characteristic D_E (diffraction coefficient with polarization E). The discussion excludes the directions of incidence and reflection $\theta = \pi \pm \theta_0$, where

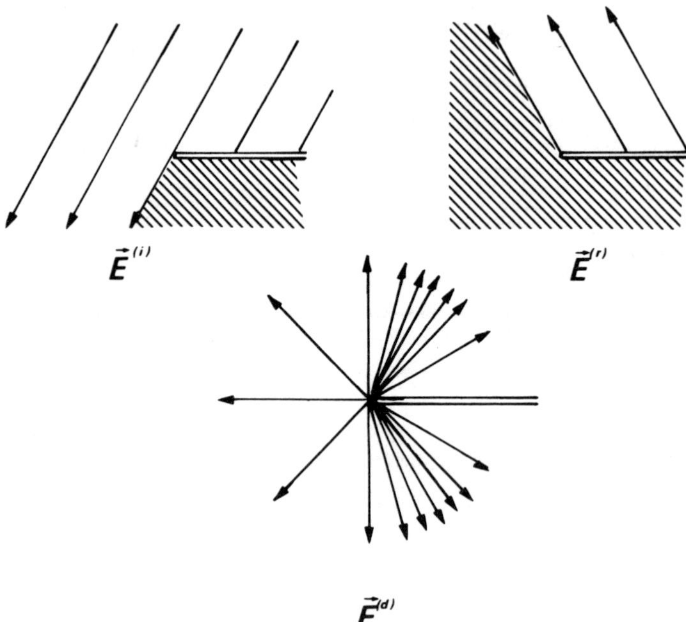

Fig. 20. *Incident waves, reflected and diffracted by the half-plane.*

the geometrical optics field is discontinuous and around which τ_1 and τ_2 can take on any finite value.

The study of the polarization H, where the magnetic field is parallel to the edge, is very similar. A correlative diffraction coefficient D_H is introduced by changing $-$ into $+$ in the brackets surrounding the second term in the expression for $H_z^{(d)}$, analogous to (15).

(c) Oblique incidence

(1) Polarization *E* (also referred to as $\perp$) The polar and semipolar coordinates (ρ, β, θ) and $(r = \rho \sin \beta,\ \theta,\ z = \rho \cos \beta)$, are introduced using notation consistent with that preceding, where $\beta = \pi/2$.

Let (β_0, θ_0) be the direction from which the incident field arrives. With polarization E, the electric field is polarized in the plane containing edge Oz of the half-plane conductor $y = 0, x > 0$; that is, reduced to its component $E_\beta^{(i)} = -E_z^{(i)}/\sin \beta_0$. The magnetic field is orthogonal to the edge ($H_z = 0$).

Component E_z of the total field allows all the others to be obtained. At a great distance, the following is found for the purely diffracted part:

$$E_z^{(d)} = D_E \frac{e^{-ik(r \sin \beta_0 - z \cos \beta_0)}}{\sqrt{r \sin \beta_0}} \tag{16}$$

which reveals the wave cones $r \sin \beta_0 - z \cos \beta_0 =$ constant, around the edge and with half-angle at the peak $(\pi/2) - \beta_0$. GTD will use rectilinear rays, inclined by $\pi - \beta_0$ to Oz, the orthogonal trajectories of the wave cones. They are reassembled as the generating lines of cones with half-angle β_0 relative to $z < 0$. Along such a ray, for example, starting at the origin, $r = s \sin \beta_0, z = -s \cos \beta_0$; hence, for an incident field where $E_s^{(i)}(0) \neq 1$:

$$E_z^{(d)} = E_z^{(i)}(0) \frac{D_E}{\sin \beta_0} \frac{e^{-iks}}{\sqrt{s}} \tag{17}$$

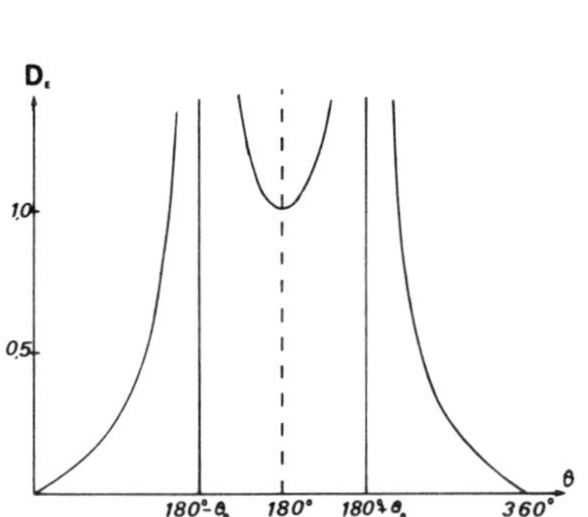

Fig. 21. *Diffraction coefficient.*

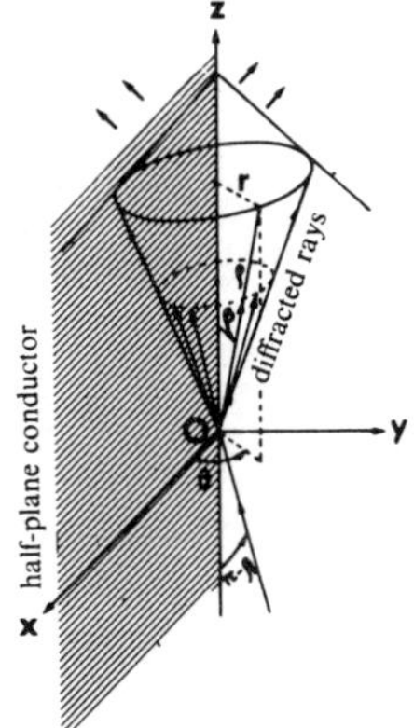

Fig. 22. *Oblique incidence onto the half-plane.*

(2) General case The polarization H (also known as $\|$), is studied in a complementary fashion. The general case is treated by resolving into the two polarizations, but the vectorial transposition of (17) is not evident. A simple rule was devised by Deschamps ([12]). Let E_1 (E_2) be the electric field corresponding to the incident (reflected) field at the origin due to the rotation $\theta - \theta_0$ ($\theta + \theta_0$) about Oz, bringing the direction of incidence (of reflection) to that of observation. Thus:

$$\vec{E}^{(d)} = [(D_H - D_E)\vec{E}_1 + (D_H + D_E)\vec{E}_2] \frac{e^{-iks}}{2 \sin \beta_0 \sqrt{s}} \tag{18}$$

C: Indirect method

1: Induced current

If the current J_S induced by the total field (see (3)) on the boundary of a conductive obstacle is known, the diffracted field results from a simple calculation of antenna radiation. The formulae (168) and (169) from Chapter 1 can be applied:

$$\vec{E}^{(d)}(x) = \frac{i}{\omega\varepsilon} (\text{grad div} + k^2) \int_S \vec{J}_S(x') G(x - x')\, dS \tag{19}$$

$$\vec{H}^{(d)}(x) = -\text{curl} \int_S \vec{J}_S(x') G(x - x')\, dS, \quad x \notin S \tag{20}$$

Without insisting on the hypotheses of regularity on the surface and knowledge of the current, which is required for an exact theory (see [51]), it should be noted that the above integral (vector potential with single layer), is continuous in the whole space and is differentiable only at points x outside S. Thus by differentiation under the summation sign:

$$\vec{H}^{(d)}(x) = -\int_S \vec{J}_S(x') \wedge \text{grad'}\, G(x - x')\, dS \tag{21}$$

In addition, if, as in Chapter 1, x tends towards a regular point x_0 on S, the limits $H^{(d)}(x_0^+)$ or $H^{(d)}(x_0^-)$ apply, depending on whether x is external or internal. These obey the boundary conditions

$$\vec{n}_0 \wedge [\vec{H}^{(i)}(x_0) + \vec{H}^{(d)}(x_0^+)] = \vec{J}_S(x_0) \tag{22}$$

$$\vec{n}_0 \wedge [\vec{H}^{(i)}(x_0) + \vec{H}^{(d)}(x_0^-)] = 0 \tag{23}$$

taking into account the continuity of the incident field across S ($\vec{n}_0$ is the normal at point x_0).

The current, linked to the field to be found, is unfortunately not known. It is clear that it satisfies the integral equations. The best method may be to use a technique very similar to that used in the standard theory of single-layer newtonian potential.

For $x = x_0$, (20) has no meaning, while the integral of (21) where G is infinite when $x' = x_0$ is improper. As in potential theory, it is demonstrated ([26]) that the integral

$$\vec{I}(x_0) = \int_S \vec{n}_0 \wedge [\vec{J}_S(x') \wedge \text{grad}' \, G(x_0 - x')] \, dS \tag{24}$$

is convergent and that

$$\vec{I}(x_0) = -\vec{n}_0 \wedge \vec{H}^{(d)}(x_0^-) - \frac{1}{2}\vec{J}_S(x_0) = -\vec{n}_0 \wedge \vec{H}^{(d)}(x_0^+) + \frac{1}{2}\vec{J}_S(x_0) \tag{25}$$

from which Maue's equation (1949) is deduced:

$$\vec{J}_S(x_0) = 2\vec{n}_0 \wedge \vec{H}^{(i)}(x_0) - 2\int_S \vec{n}_0 \wedge [\vec{J}_S(x') \wedge \text{grad}' \, G(x_0 - x')] \, dS \tag{26}$$

an integral equation of the second type which rigorously solves the current problem. Once this has been found, the diffracted field is calculated using (19) and (20). Present numerical methods have meant that equation (26) has taken on an importance that its rather daunting appearance would not have suggested.

Notes

- Note that (25) is the mean:

$$\vec{I}(x_0) = -\vec{n}_0 \wedge \frac{\vec{H}^{(d)}(x_0^+) + \vec{H}^{(d)}(x_0^-)}{2}$$

- The above formulation, suitable for direct numerical calculation, does not involve the use of the finite parts of divergent integrals or distributions. In these, the fields are only involved by virtue of their integrability and not of their point-values. On the contrary, if the starting point had been the electric field (19), the double differentiation under the summation sign would have given a divergent integral. In addition, even with the use of a finite part, the equation resulting would have been of the first kind.
- If (26) could be solved by successive approximations, the first would be obtained by taking a zero current in the integral. The following would remain:

$$\vec{J}_S(x_0) = 2\vec{n}_0 \wedge \vec{H}^{(i)}(x_0) \tag{27}$$

and this will be encountered again in the physical optics approximation.

2: Physical optics approximation

This consists of calculating the diffracted field exactly (19), (20), starting with the approximate current found by geometrical optics reasoning. Consider the case of a convex body on the surface S of which a closed curve C separates the illuminated part S_0, in the sense used in geometrical optics, from the obscured part S_1. At all regular points of S_0, the tangential plane is treated as a perfect reflector. So, according to (106) of Chapter 2, the total magnetic field is:

$$\vec{H} = \vec{H}^{(i)} + \underline{\vec{H}}^{(i)} = 2\vec{H}^{(i)}_{\text{tang}}. \tag{28}$$

and the current is:

$$\vec{J}_S = \vec{n} \wedge \vec{H} = 2\vec{n} \wedge \vec{H}^{(i)} \tag{29}$$

In a similar way, the density of charge can be found: $\rho_S = 2\varepsilon\vec{n}\cdot\vec{E}^{(i)}$. The physical optics approximation is based on taking in (19) and (20):

$$\vec{J}_S = 2(\vec{n} \wedge \vec{H}^{(i)})\mathbf{1}_{S_0} \tag{30}$$

hence, explicitly, the total field:

$$\vec{E} = \vec{E}^{(i)} - \frac{2}{\mathrm{i}\omega\varepsilon}(\text{grad div} + k^2)\int_{S_0}(\vec{n} \wedge \vec{H}^{(i)})G\,dS \tag{31}$$

$$\vec{H} = \vec{H}^{(i)} - 2\,\text{curl}\int_{S_0}(\vec{n} \wedge \vec{H}^{(i)})G\,dS \tag{32}$$

From a practical point of view, it should be noted that this procedure gives acceptable results in the cases specified in the previous chapter: small λ relative to the dimensions of the obstacle and the principal radii of curvature, with a relatively distant observation point. It is generally admitted that the approximation improves on that of geometrical optics, but this is not always the case. On the other hand, there are cases in which the latter cannot be used, and where the former can give good results (small- or large-angle cones).

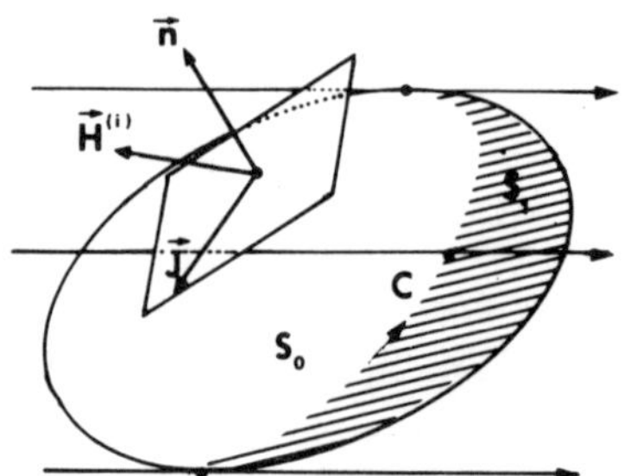

Fig. 23. *Physical optics approximation.*

From a theoretical point of view the method has an obvious drawback: it does not give the initial boundary conditions. In addition, the discontinuity of current across the curve C—which is introduced arbitrarily—does not accord with the observed phenomena (creeping waves, for example). Finally, if, in order to use expressions suited to numerical calculation, the above integrals are differentiated under care summation signs without sufficient care, this discontinuity violates the conservation of electric charge. For this reason, it is instructive to treat this problem directly (see next section).

3: Radiating cells

'Radiating cell' is the name given here to a limited element with a regular surface S, carrying a given distribution of electric and magnetic currents. This is encountered in many antenna problems. The cell is oriented by its normal n, and the geodesic normal g along its border-line is oriented towards its exterior. This is then oriented by its tangent t, in such a way as to make the frame (t, n, g) right-handed.

As in Chapter 1, the bicomplex current will be used:

$$\vec{K}_S = \eta \vec{J}_S + \vec{M}_S$$

The radiation from it (Chapter 1, equation (146)) is:

$$\vec{C} = \left[\frac{i}{k}(\text{grad div} + k^2) - j \text{ curl}\right](\vec{\mathbf{K}}_S * G) \tag{33}$$

that is, explicitly:

$$\vec{C}(x) = \frac{i}{k}(\text{grad div} + k^2)\int_S \vec{K}_S(x')G(x - x')\, dS - j \text{ curl}\int_S \vec{K}_S(x')G(x - x')\, dS \tag{34}$$

a compact representation of the usual fields:

$$\begin{aligned}\vec{E} &= -\frac{1}{i\omega\varepsilon}(\text{grad div} + k^2)\int_S \vec{J}_S G\, dS + \text{curl}\int_S \vec{M}_S G\, dS \\ \vec{H} &= -\frac{1}{i\omega\mu}(\text{grad div} + k^2)\int_S \vec{M}_S G\, dS - \text{curl}\int_S \vec{J}_S G\, dS\end{aligned} \tag{35}$$

These fundamental expressions obey Maxwell's equations, the radiation conditions and the boundary conditions:

$$\vec{K}_S = -j\vec{n} \wedge [\vec{C}]_S \tag{36}$$

They also have the advantage of including the charge effect:

$$\chi_S = \frac{1}{v}\vec{n}\cdot[\vec{C}]_S$$

Their drawback, from a practical point of view, is that they are less easy to handle than the direct expressions obtained by differentiation under the summation sign. By differentiation with respect to x and use of some common identities, the following would be found:

$$\vec{C} = \frac{i}{k}\int_S [(\vec{K}_S\cdot \text{grad})\text{ grad } G + k^2 G + ijk \text{ grad } G \wedge K_S]\, dS \tag{37}$$

The current can be replaced by expression (36). Only those examples where the field is assumed to be zero on the negative side of S will be considered here. Using simple but lengthy standard vector analysis transformations, the expression below is established:

$$\vec{C} = -\int_S [(\vec{n}\cdot\vec{C})\,\text{grad}'\,G + ijk(\vec{n}\wedge\vec{C})G + (\vec{n}\wedge\vec{C})\wedge\text{grad}'\,G]\,dS + \frac{1}{ijk}\int_C \text{grad}'\,G(\vec{C}\cdot\vec{t})\,ds \qquad (38)$$

where the field on the positive side of S appears in the surface integral. The validity of this approach is confirmed particularly in the case of closed surfaces where, in addition, the last term disappears. Thus equation (189) of Chapter 1 is re-established, taking account of the fact that grad' $G = -$grad G.

The contour integral represents radiation from a linear charge distribution along side C. These charges ensure the conservation of electricity and magnetism on the cell. Their role was noted by Kottler (1923).

Note that a different curvilinear integral appears in Kirchhoff's representation (Appendix 3).

In the far field, the last integral tends to reduce to just the radial term (equation (192) of Chapter 1). It can be considered as the correction that ensures, in conformity with Maxwell's equations, the transversality of the field at infinity. Figuratively, it could be said that a radiating cell equipped with its Kottler charges has an autonomous electromagnetic existence. In antenna calculations involving the breakdown of a surface into adjacent cells, the charges on the arcs common to two of them are neutralized (see Stokes's theorem). They disappear on closed surfaces.

4: Applications

(a) Physical optics approximation

The illuminated part S_0 of a diffracting obstacle is considered as a radiating cell. From (38) the following can be deduced, in a similar way to (31) and (32):

$$\vec{E} = \vec{E}^{(i)} - 2\int_{S_0} [(\vec{n}\cdot\vec{E}^{(i)})\,\text{grad}'\,G - i\omega\mu(\vec{n}\wedge\vec{H}^{(i)})G]\,dS + \frac{2}{i\omega\varepsilon}\int_c \text{grad}'\,G(\vec{H}^{(i)}\cdot\vec{E})\,ds \qquad (39)$$

$$\vec{H} = \vec{H}^{(i)} - 2\int_{S_0} (\vec{n}\wedge\vec{H}^{(i)})\wedge\text{grad}'\,G\,dS \qquad (40)$$

The formulae are used, in particular, in approximate calculations for reflectors.

(b) Return to Huygens' principle

The formulae in equation (203) of Chapter 1 (where grad G is changed to $-$grad' G) express any e.m. field outside a closed surface S by means of its values on the surface. Cutting out an open cell on S has no physical meaning in itself, but it can be a useful calculational intermediary (cf Kirchhoff's aperture theory).

The above formulae would give fields that obey Maxwell's equations. If S is a cell with contour C, the electric part of (38) is given by:

$$\vec{E} = -\int_S [(\vec{n}\cdot\vec{E})\,\mathrm{grad'}\,G - i\omega\mu(\vec{n}\wedge\vec{H})G + (\vec{n}\wedge\vec{E})\wedge\mathrm{grad'}\,G]\,dS + \frac{1}{i\omega\varepsilon}\int_c (\vec{H}\cdot\vec{t})\,\mathrm{grad'}\,G\,ds \quad (41)$$

The correlative $E \to \eta H, \eta H \to -E, \varepsilon \to \mu$ allows the expression of the magnetic field to be established. Using the calculation on page 87, (41) is equivalent to:

$$\vec{E} = -\frac{1}{i\omega\varepsilon}(\mathrm{grad\,div} + k^2)\int_S (\vec{n}\wedge\vec{H})G\,dS - \mathrm{curl}\int_S (\vec{n}\wedge\vec{E})G\,dS \quad (42)$$

(c) The far field

These are reduced to the transverse part of the second term in (38). There is no point in recalculating, as this would give equation (194) of Chapter 1 again. In the physical optics approximation, (39), the following would be found:

$$\vec{E} = \vec{E}^{(i)} + 2ik\eta G_0\vec{u}\wedge\int_{S_0} (\vec{n}\wedge\vec{H}^{(i)})\wedge\vec{u}\,e^{i\vec{k}\cdot\vec{x}'}\,dS \quad (43)$$

and for the Huygens cell (41):

$$\vec{E} = ikG_0\vec{u}\wedge\int_S [\eta(\vec{n}\wedge\vec{H})\wedge\vec{u} + \vec{n}\wedge\vec{E}]\,e^{i\vec{k}\cdot\vec{x}'}\,dS \quad (44)$$

where $G_0 = G(x)$ is referred to the origin.

As in Chapter 1, it will be noted that the magnetic field is given by: $(u\wedge E)/\eta$. These results are constantly in use in antenna calculations.

D: Diffraction through apertures

1: The aperture problem

Diffraction through apertures in a screen is encountered in many antenna problems concerning microwaves. Some antennas are effectively radiating apertures: holes or slots in a guide or cavity wall, open guides, horns. Others, such as lenses and reflectors, can be treated using a model, more or less equivalent, of an aperture in a plane screen. A screen will be a sheet, of a particular thickness and generally conductive, separating a half-space Ω^-, containing the sources Σ, from a half-space Ω^+, bounded by faces S^- and S^+. The apertures S_0^- and S_0^+ bounded by C^- and C^+, leave the remaining areas S_1^- and S_1^+. These contours are oriented as explained above.

The problem of the conductive screen is, at least in theory, solvable using Maue's equation (26). In practice, however, approximate methods are used.

The physical optics approximation consists of treating the single side S_1^-, to which the integrals of (31) and (32) can simply be extended, as though it were illuminated. In fact, physical optics proper uses Kirchhoff's approximation, which consists of calculating in Ω^+ the field created by the radiating cell S_0^+, cut out of the incident wave by the obscured side of the screen. In other words, starting with a Huygens representation in Ω^+ of the incident field, derived from its values on the geometric surface S^+, the diffracted field is considered as being created only by the secondary sources in part S_0^+. This approximation has the drawback in e.m. theory of not taking any account of the physical nature of the screen.

It is interesting to discuss the methods indicated for the specific but important case of a very thin plane screen. This case has given rise to a great deal of research, which cannot be summed up here (see [8]). A few simple calculations will, however, be given, but first the important principle of images will be discussed.

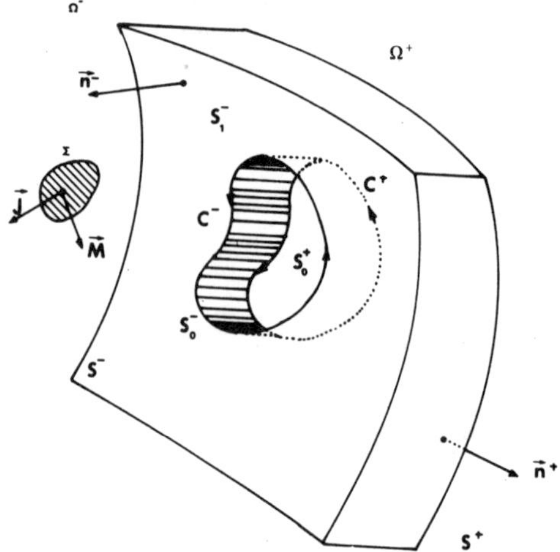

Fig. 24. *Diffraction by an aperture.*

2: Image principle

This is constantly used in antenna calculations above ground planes, and is therefore summarized here.

Consider a plane S separating two half-spaces, one of which contains the sources $\Sigma(J, M)$ with radiation $E^{(i)}$, $H^{(i)}$, in free space, given by equation (160) of Chapter 1. The point symmetric to a point (or vector) x, relative to the plane S, will be denoted $\underline{x}$. The following fictitious sources are introduced into the empty-half space:

$$\vec{J}(\underline{x}) = -\underline{\vec{J}}(x), \quad \vec{M}(\underline{x}) = \underline{\vec{M}}(x) \tag{45}$$

(see Chapter 2, equation (125)). These sources are called image-sources.

The parity, relative to S, of the vector differentiations grad and div, and the imparity of curl (cf the outmoded distinction between axial and polar vectors) is sufficient to show that the field $E^{(r)}$, $H^{(r)}$ (see Chapter 1, equation (160)), created by the image sources, obeys:

$$\vec{E}^{(r)}(\underline{x}) = -\underline{\vec{E}}^{(i)}(x), \quad \vec{H}^{(r)}(\underline{x}) = \underline{\vec{H}}^{(i)}(x) \tag{46}$$

From this can be found a total field, created by the set of sources Σ and the image sources:

$$\vec{E}(x) = \vec{E}^{(i)}(x) - \underline{\vec{E}}^{(i)}(\underline{x}), \quad \vec{H} = \vec{H}^{(i)}(x) + \underline{\vec{H}}^{(i)}(\underline{x}) \tag{47}$$

reduced at all points x_0 in S to:

$$\vec{E}(x_0) = 2\vec{E}_n^{(i)}(x_0), \quad \vec{H}(x_0) = 2\vec{H}_S(x_0) \tag{48}$$

such that, if surface S were conductive, the usual boundary conditions would be satisfied. The uniqueness of the radiated fields allows the following extension of the standard principle of electric images to be stated: the field (J, M) radiated by an antenna above a perfect plane reflector is the same as that created in the half-space envisaged by the antenna and its image, in the sense of (45), in the absence of a reflector.

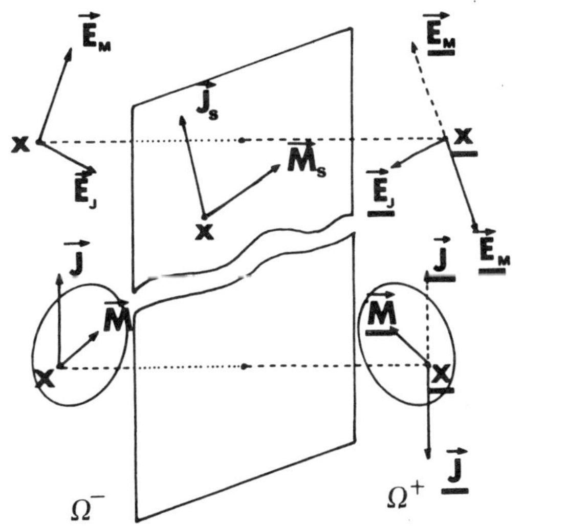

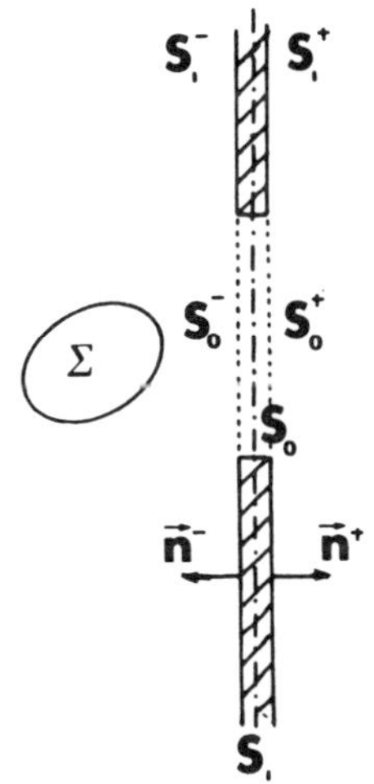

Fig. 25. *Plane sources and image sources.*

Fig. 26. *Thin plane screen.*

Exercise

Using the notation:

$$\underline{\vec{K}} = \eta\underline{\vec{J}} - j\underline{\vec{M}}, \quad \underline{\vec{C}} = \underline{\vec{E}} - j\eta\underline{\vec{H}}$$

the most compact expression for the symmetry relations can be obtained:

$$\vec{K}(\underline{x}) = -{}_{(S)}\vec{K}(x), \quad \vec{C}^{(r)}(\underline{x}) = -\underline{\vec{C}}^{(i)}(x), \quad \vec{C}(x) = \vec{C}^{(i)}(x) - \underline{\vec{C}}^{(i)}(\underline{x}).$$

3: Thin plane screen

(a) Notation

The following abbreviated notation will be used in the following section:

$$\Phi(A, S) = \frac{1}{ik} (\text{grad div} + k^2) \int_S (\vec{n} \wedge \vec{A}) G \, dS \tag{49}$$

$$\psi(A, S) = \text{curl} \int_S (\vec{n} \wedge \vec{A}) G \, dS$$

The normal is external to the domain of which S is all or part of the boundary. In addition, the formula will be more symmetrical if ηH is taken as the measurement of the magnetic field (that is, in V/m).

(b) Physical optics approximation

Only the electric field will be calculated here. S_1^- is the conductive, illuminated part of the screen, according to (31):

$$\vec{E} = \vec{E}^{(i)} - 2\Phi(H^{(i)} \cdot S_1^-) \tag{50}$$

Note

Huygens' representation (42) of the incident field by its values on the plane of the screen is expressed here, in abbreviated form:

$$1_{\Omega^+} \vec{E}^{(i)} = -\Phi(H^{(i)}, S^+) - \psi(E^{(i)}, S^+) \tag{51}$$

The image-sources (45) relative to this plane create:

$$1_{\Omega^-} \vec{E}^{(r)} = -\Phi(H^{(r)}, S^-) - \Psi(E^{(r)}, S^-) = \Phi(H^{(i)}, S^+) - \Psi(E^{(i)}, S^+) \tag{52}$$

taking account of (46) and the orientation of the normals. In Ω^+, therefore, by addition and subtraction of (51) and (52):

$$\vec{E}^{(i)} = -2\Phi(H^{(i)}, S^+) = -2\Psi(E^{(i)}, S^+) \tag{53}$$

(c) Severin's formulae (1952)

If (53) is introduced into (50), then:

$$\vec{E} = -2\Phi(H^{(i)}, S_0^+) \tag{54}$$

Likewise $-2\psi(H^{(i)}, S_0^+)$ will be obtained for H. By returning to explicit expressions:

$$\left.\begin{aligned} \vec{E} &= -\frac{2}{i\omega\varepsilon} (\text{grad div} + k^2) \int_{S_0^+} (n \wedge H^{(i)}) G \, dS \\ \vec{H} &= -2 \, \text{curl} \int_{S_0^+} (n \wedge H^{(i)}) G \, dS \end{aligned}\right\} \tag{55}$$

The Severin formulae are simpler than (50) and the similar expression for H, to which they are equivalent in Ω^+, and are extended to an aperture assumed to be limited. In addition, they give results that are closer to those observed than the Kirchhoff approximation (see below).

Note
In a similar way, in Ω^-:

$$\vec{E} = \vec{E}_0 + 2\Phi(H^{(i)}, S_0^-) \tag{56}$$

(d) Kirchhoff's approximation
The physical optics approximation starts with true electric currents approximated on the illuminated surface of the conductive screen. That of Kirchhoff uses fictitious electric and magnetic currents on the incident part of the wave cut out by the aperture. The basic hypothesis is that the screen does not disturb this wave. This is more open to debate when the screen is more conductive and the aperture smaller. (42) can simply be applied. In Ω^+:

$$\vec{E} = -\Phi(H^{(i)}, S_0^+) - \Psi(E^{(i)}, S_0^+) \tag{57}$$

Taking account of (51), however:

$$\vec{E} = \vec{E}^{(i)} - \Phi(H^{(i)}, S_1^-) - \Psi(E^{(i)}, S_1^-) = \vec{E}^{(i)} + \vec{E}^{(d)} \tag{58}$$

The diffracted part is created by the electric and magnetic currents, respectively $n^- \wedge H^{(i)}$ and $-n^- \wedge E^{(i)}$, induced by the incident field on the illuminated face. According to the hypothesis, they are zero on the obscured face. In fact, the case of a perfectly absorbent screen is treated. For conductive screens, the Severin formulae are better ([41], page 119).

Note
The approximations discussed above do not give good results close to the edge; this result might have been anticipated. Braunbeck (1950) improved them, by using the current given by the exact half-plane theory.

4: Babinet's principle

(a) Definitions
In the thin plane screen described in the previous paragraph, (S_1: conductor; S_0: aperture) the full screen ($S_1 \cup S_0$ conductor) and the complementary screen (S_0: conductor; S_1: vacuum) can be distinguished.

Associated with the sources Σ situated in Ω^- are the correlative sources Σ' defined on page 19, creating the correlative field of Σ. (Example: Hertz's doublet and elementary current loop, dipole antennas and slots, plane wave rotated by 90° about the direction of propagation, etc).

In classical optics, Babinet's principle relates the diffractions by means of complementary screens. It can be extended to electromagnetism.

(b) Diffraction by plane screen
The incident field $C^{(i)}$ created by the sources Σ gives a total field $C = C^{(i)} + C^{(d)}$,

the diffracted part of which is the radiation from plane currents on S_1. Hence the symmetry relationship (see Exercise on page 91):

$$\vec{C}^{(d)}(\underline{x}) = \underline{\vec{C}}^{(d)}(x)$$

From this it is deduced that the diffracted electric field is tangential and the magnetic field is normal at any point in the plane $S_1 \cup S_0$. On S_0, therefore:

$$E_n^{(d)} = 0, \quad H_s^{(d)} = 0 \tag{59}$$

completed by the boundary conditions on conductor S_1:

$$E_s^{(d)} = -E_s^{(i)}, \quad H_n^{(d)} = -H_n^{(i)} \tag{60}$$

Note

In the presence of the entirely conductive screen, the sources Σ would create the following field (see page 91):

$$\vec{C}_0 = (\vec{C}^{(i)} - \underline{\vec{C}}^{(i)})1_{\Omega^-} \tag{61}$$

(c) Diffraction by the complementary screen

The correlative incident field

$$\vec{C}'^{(i)} = -j\vec{C}^{(i)} \tag{62}$$

created by the correlative sources would produce, in the presence of the full screen, a field C'_0 similar to (61). Consider *a priori* the following field:

$$\vec{C}' = j\vec{C}^{(d)} \qquad \text{in} \quad \Omega^+ \tag{63}$$

$$\vec{C}' = \vec{C}'_0 - j\vec{C}^{(d)} \qquad \text{in} \quad \Omega^- \tag{64}$$

It satisfies Maxwell's equations everywhere. It is clear that it is continuous across S_1 and satisfies the boundary conditions of the perfect conductors on S_0, taking account of (59) and (60), and it can even be shown that Meixner's conditions (page 22) apply on the edges if they are in the first diffraction. Finally, the diffracted part satisfies the radiation condition. Field (63), (64) is therefore the single total field for the diffraction of field (62) by the complementary screen, throughout the whole space.

Note

In a similar way, also, in Ω^+:

$$\vec{C} = -j\vec{C}'^{(d)} \tag{65}$$

(d) Babinet's principle

This can be seen in (63) and (65). In a complementary manner, Babinet's principle exchanges the total field in each diffraction with the diffracted part of the other in the region that does not contain the sources.

Notes

- In Ω^- there is a similar rule for the field $C - C_0$:

$$\begin{aligned} \vec{C}'_0 - \vec{C}' &= j\vec{C}^{(d)} \\ \vec{C}_0 - \vec{C} &= -j\vec{C}'^{(d)} \end{aligned} \tag{66}$$

- In Ω^+:

$$\vec{C}^{(i)} = \vec{C} - \vec{C}^{(d)} = \vec{C} + j\vec{C}' \tag{67}$$

hence the explicit formulation of the principle:

$$\vec{E}^{(i)} = \vec{E} - \eta\vec{H}', \quad \vec{H}^{(i)} = \frac{1}{\eta}\vec{E}' + \vec{H} \tag{68}$$

B: Geometrical theory of diffraction (GTD)

1: Principles ([12]), [17], [18], [22])

GTD was first developed about twenty years ago, in an extension of work by the Soviet school, carried out by Keller. The theory became a great success, and this can be accounted for by its intuitive principles and its practical applications, in particular for composite obstacles. As an approximation of high-frequency diffraction by extension of geometrical optics, GTD encounters difficulties with the areas around caustics, shadow boundaries, sources, etc. Successive refinements to the available methods have been made to this more or less empirical method, the basic ideas of which are explained here.

The first is that diffracted fields are transported along the rays (diffracted rays) in accordance with the laws of geometrical optics (see below). It was suggested by the theories of the cylinder (page 78) and half-plane (page 80) and was already partially developed in the traditional treatment of Young's slits.

Figure 27 shows the example of an opaque convex cone, illuminated by a point source P. It is supposed that at an observation point Q other rays than those of standard optics converge. Thus the paths PS, PK, and PIJ illustrate diffraction by, respectively, the apex S, the base contour and the curved surface.

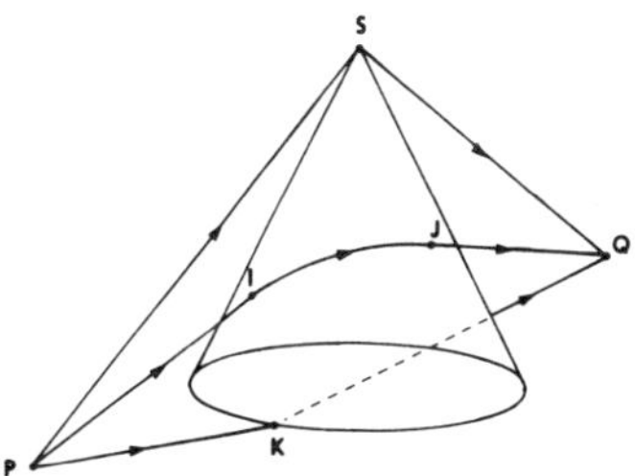

Fig. 27. *Diffracted rays.*

The concept of a diffraction coefficient completes the notion of the diffracted rays. These coefficients ensure the connection of fields at the points of the obstacles from which the diffracted rays issue. Diffracted rays and diffraction coefficients are proper to GTD, and enter into the standard calculations of differential geometry and geometrical optics.

2: Construction of rays

Figure 27 shows how new relations describing optical paths are found. A ray can be made to meet an apex, a contour, or touch a convex surface. In more complicated problems a single ray can contain several links. It is postulated that such rays continue to obey Fermat's principle (page 62): any ray PQ is stationary with respect to the varied paths PQ constrained by the same relationship. For example, the unique ray PSQ in the diagram is formed by two rectilinear rays PS and SQ.

The idea of extending Fermat's principle is intuitive. In fact, it extrapolates to high frequencies the results already established, in cases for which an exact solution can be found (canonical cases), such as the cylinder of revolution or the half-sheet.

3: Diffraction coefficients

This is the most difficult problem (in GTD). On a path parametrized by arc s, the transport of the field is reduced to that of scalars $u(s)$ (components, norm, or intensity, phase). At the origin $s = s_0$ of a diffracted portion, there is no reason for such a scalar to be continuous (cf the reflection or transmission coefficient). A scalar diffraction coefficient D, is introduced and has the significance of a transmission coefficient:

$$u(s_0^+) = Du(s_0^-) \tag{69}$$

At the high frequencies, diffraction by an obstacle can be considered to be determined by the only field in the vicinity of the obstacle (Fock's local field principle). If there is a canonical problem for a case 'tangential' to the obstacle considered, the fields in both neighbourhoods are identified and the coefficient D, determined by the canonical problem, is adopted as a rule.

Note that the physical reality is in no way described at a so-called point of diffraction. This is a mathematical artifice, which gives the desired field $u(s)$ only asymptotically for $k(s - s_0) \gg 1$. In fact, the results remain acceptable to within a few wavelengths of s_0 (sometimes even one or two).

4: Diffraction by an edge

(a) Rays

Any stationary path PQ meeting a contour C is formed of two rectilinear rays PK and KQ. For a similar varied path $PK'Q$ defined on C by: $\vec{KK'} = t\, ds$, it follows that:

$$dr = ds \cos \beta, \quad dr' = -ds \cos \beta$$

or:

$$r = |\vec{PK}|, \quad r' = |\vec{KQ}|, \quad \beta = (\vec{t}, \vec{PK}), \quad \beta' = (\vec{t}', \vec{KQ}) \tag{70}$$

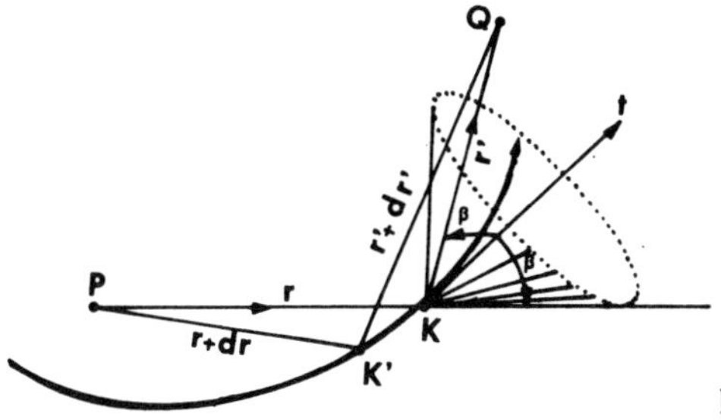

Fig. 28. *Diffraction by an edge.*

The stationarity condition $dr + dr' = 0$ leads to $\beta = \beta'$. This result was encountered, with different notation, for the rectilinear edge of the half-plane. An image can be formed of a stretched wire PQ, passing without friction through a small ring, sliding without friction on the rigid curve C. An infinite number of diffracted rays will correspond to a given incident ray PK, along the generating lines of the half-cone of revolution about the tangent at K and the half-angle at the apex β.

If the incident ray is normal to the contour ($\beta = \pi/2$), the diffracted rays generate the plane normal to C in K.

Application: Normal incidence on a plane aperture.
When P is at infinity in the direction normal to the plane of contour C, the planes normal to C envelop the cylinder, which has as its straight section the evolute of C. All the diffracted rays meet the contour C and are tangential to this cylinder. Together they form the caustic for the congruence of the diffracted rays.

Example: Circular aperture. The caustic is reduced to the circle C and its axis. It would seem that only the two diffracted rays PKQ and $PK'Q$ need to be considered, in the meridian plane that contains an image point Q. In fact, among the rays diffracted at K there is, in addition to KQ, the diameter KK', which gives rise to further diffraction at K' and so on. There is, therefore, a double infinity of diffracted paths, PKQ, $PKK'Q$, $PKK'KQ$, ..., $PK'Q$, $PK'KQ$, etc.

Note
Figure 29b equally represents diffraction by a rectilinear slot.

Exercise
On any diffracted ray issuing from K, on any contour C, K is a focal point. Determine the other focal point, F.

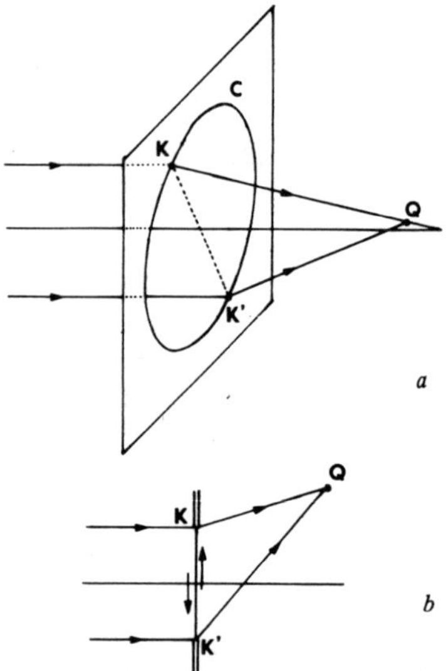

Fig. 29. *Circular aperture.*

Hint

Plot the vector $KF = \vec{u}$ with Serret-Frenet's frame at K and express the fact that the absolute displacement dF for a displacement dK on C is parallel to the unit vector $\vec{u}$. With Frenet's formulae:

$$\frac{1}{\rho} = \frac{\beta'(s)}{\sin \beta} + \frac{\cos \delta}{R \sin^2 \beta} \tag{71}$$

where R designates the radius of curvature (>0) of C at K, and δ the angle of the ray with the principal normal. $\beta(s)$ is determined by the incident ray and the chosen diffracted ray by β and δ.

(b) Diffraction coefficient

In the following, P will be assumed to be at infinity. Let C be the edge of a conductive surface. Calculation is based on a field component $u(s)$ along a diffracted ray KQ. It is convenient to take K as the origin: $s = 0$. Thus, according to the formula (s. 122):

$$u^{(d)}(s) = v(O)\sqrt{\frac{\rho}{s(s+\rho)}}\, e^{-iks} \tag{72}$$

where $v(O) = \lim_{s \to 0} u^{(d)}(s)\sqrt{s}$ and ρ is given by (71).

The coefficient $v(O)$ is determined using the appropriate canonical case, constituted by Sommerfeld's conductive half-plane, tangential at K. For this the second focus F of the diffracted ray is projected to infinity and (72) is reduced to:

$$u^{(d)}(s) = v(O)\frac{e^{-iks}}{\sqrt{s}} \tag{73}$$

By coming closer to (17), a formula of the following type can be obtained:

$$u^{(d)}(s) = u^{(i)}(O)\,\frac{D}{\sin\beta}\sqrt{\frac{\rho}{s(s+\rho)}}\,e^{-iks} \tag{74}$$

and can be extended to the vector fields by means of a matrix diffraction coefficient, or in a similar way to (18) ([12], [16], [22]).

Exercises:

- Diffraction of a plane wave by a rectilinear slot ([18]).
- Diffraction, for normal incidence, of a plane wave by a circular aperture ([10]).

5: Diffraction by a convex body

(a) Rays

(1) Cylinder This is based on the study of the cylinder of revolution. Consider a convex cylinder illuminated by a linear source parallel to the generating lines. In a plane section, a path PQ is formed by two rectilinear rays PK_1 and K_2Q tangential to an arc K_1K_2 of section C. Ray PK_1 is incident, the arc K_1K_2 corresponds to a creeping wave (see page 79), and the emergent ray K_2Q is the diffracted ray. When only a convex finite cylindrical section is considered, there is a second path $PK'_1K'_2Q$. The set of diffracted rays is made up of that of the half-tangents to contour C. An image point Q is, therefore, the end of an infinite number of paths comprising, in opposite directions, from K_2 or K'_2 any number of paths along the contour. In fact, the attenuation of the field along the creeping rays is such that in practice only the direct paths need be considered.

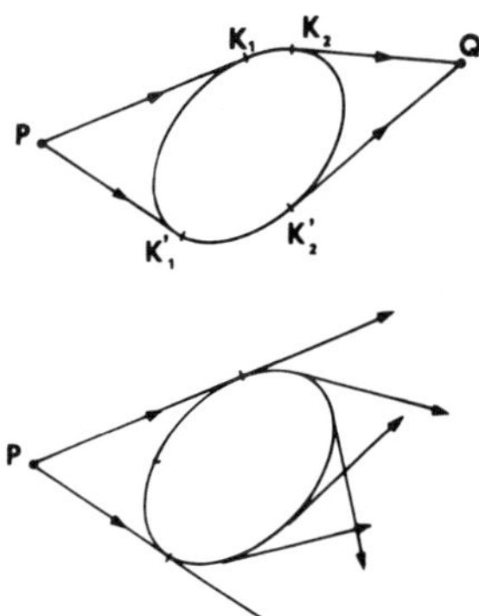

Fig. 30. *Diffraction by a convex cylinder.*

Bodies of general form A stationary path between P and Q passes along the surface S of the obstacle, and is made up of two rectilinear rays tangential to S, PK_1 incident, and K_2Q diffracted, linked with an arc of geodesic K_1K_2 (the picture of a wire stretched between P and Q, touching S without friction). The source P gives rise to a congruence of the diffracted rays obtained in the following way. For each incident ray tangential to S there is a corresponding geodesic, tangential to it at the point of contact. On S, therefore, there is a nappe of geodesics formed by the extension of the cone generating lines from the apex P, circumscribed at S. The set of the half-tangents to these geodesics with the same orientation forms the congruence of the rays. S is a sheet of the caustic.

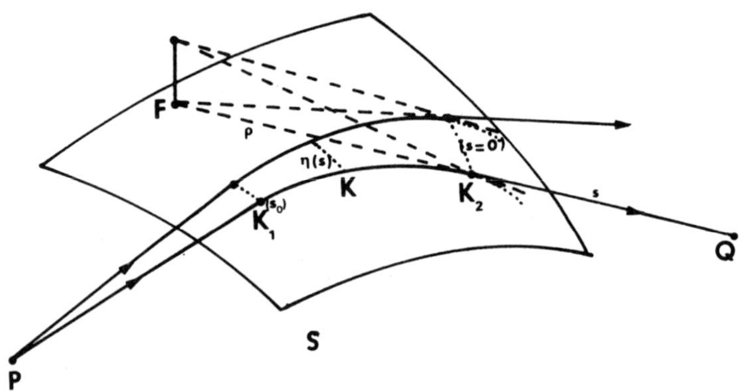

Fig. 31. *Diffraction by a convex body.*

It is helpful to parametrize each path by the arc t of the geodesics ($t = 0$ at K_1, $t = t_0$ at K_2), so as to conserve, as above, s on the diffracted rectilinear rays ($s = 0$ at K_2). Consider on S, an elementary (that is narrow) layer of geodesics, arising from the same source P. Considerations of differential geometry determine the width $\omega(t)$ of the layer and the position of the second focus $F(FK_2 = \rho)$, on the diffracted ray issuing from K_2.

(b) Field transport

Let $u(t)$ be a component of the field. Geometrical optics gives the incident field at K_1, $u_0^{(i)}$. The power transported by this component along an elementary layer is proportional to $P(t) = |u(t)|^2 w(t)$ and decreases as $\exp[-2A(t)]$—there is continuous emission by the tangential diffracted rays—where, according to a known law, A is the integral of O at t of the attenuation coefficient $\alpha(t)$. So along a creeping ray:

$$u(t) = D_{K_1} u_0^{(i)} \sqrt{\frac{w(0)}{w(t)}}\, e^{-A(t) - ikt} \tag{75}$$

Then along the ray issuing from K_2, one takes the following, as in (74):

$$u^{(d)}(s) = D_{K_2} u(t_0) \sqrt{\frac{\rho}{s(s+\rho)}}\, e^{-iks} \tag{76}$$

By reciprocity $D_{K_1} = D_{K_2}$. These coefficients are identified with those arising in canonical cases, such as the cylinder of revolution or spheres osculating to the obstacle considered. For example, in the case of a cylinder with finite section and perimeter l, if the incidence is normal, the results (9) will be used.

Notes

• The geometric, rather than physical, significance of each of the creeping waves has already been mentioned (what is the significance of a field that 'circulates' indefinitely around a cylinder?). In addition, a component tangential to the electric field linked to a creeping ray cannot represent a real field. Artifices, boundary layers, and joining coefficients have been proposed for connecting them.

• In the same way, various corrections of caustics, of shadow boundaries, etc, have improved the initial theory. These ideas have been inspired by canonical cases or other approximate methods (such as physical optics) but do not always agree with the use of transition zones with zero corrective factors in cases where the geometric field is infinite.

Bibliography to Part 1

[1] ABRAMOWITZ M. and STEGUN I. A. *Handbook of Mathematical Functions*, Nat. Bur. of Standards, Washington, 1970.

[2] BAKER B. B. and COPSON E. T. *The Mathematical Theory of Huygens' Principle*, Oxford, Clarendon Press, 1950.

[3] BATEMAN H. *The Mathematical Analysis of Electrical and Optical Wave Motion*, Cambridge University Press, 1914.

[4] BATEMAN H. *Partial Differential Equations of Mathematical Physics*, Cambridge University Press, 1932.

[5] BECHTEL M. E. *Application of geometric diffraction theory to scattering from cones and disks*, Proc. I.E.E.E., **53**, August 1965. p. 879.

[6] BORN M. and WOLF E. *Principles of Optics*, Pergamon, London, 1964.

[7] BOUIX M. *Les discontinuités du rayonnement électromagnétique*, Dunod, Paris, 1966.

[8] BOUWKAMP C. J. *Diffraction theory*, Reports on Progress in Physics, **XVII**, 1954, London, p. 35.

[9] BREMMER H. *Propagation of Electromagnetic Waves*, Handbuch der Physik, Vol. **XVI**, Springer, Berlin, 1958, p. 423–639.

[10] CHEROT T. E. *Calculation of Near Field of Circular Aperture Antenna Using Geometrical Theory of Diffraction*, I.E.E.E. Trans., E.M.C., **13**, May 1971, p. 29.

[11] COLLIN R. E. and ZUCKER F. J. *Antenna Theory*, McGraw-Hill, New York, 1969.

[12] DESCHAMPS G. A. *Ray techniques in electromagnetics*. Proc. I.E.E.E., **60**, September 1972, p. 1022.

[13] FOURNET G. *Electromagnétisme*, Masson, Paris, 1979.

[14] JESSEL M. *Contribution aux théories du principe de Huygens et de la diffraction*, Publ. scient. du ministère de l'Air, Paris, 1963.

[15] JONES D. S. *The Theory of Electromagnetism*, Pergamon, London, 1964.
[16] KELLER J. B. *Diffraction by an Aperture*. J. of Appl. Physics, **28**, April 1957, p. 426; May 1957, p. 570.
[17] KELLER J. B. *A Geometrical Theory of Diffraction*, Proc. Symp. of Appl. Math., **8**, McGraw-Hill, New York, 1958.
[18] KELLER J. B. *Geometrical Theory of Diffraction*, J. Optical Soc. of America, **52**, February 1962, p. 116.
[19] KLINE M. and KAY I. K. *Electromagnetic Theory and Geometrical Optics*, Inter-science Publ., New York, 1965.
[20] KNOTT E. F. and SENIOR T. B. A. *Comparison of three high-frequency diffraction techniques*, Proc. I.E.E.E., **62**, November 1974, p. 1468.
[21] KOTTLER F. *Diffraction at a black screen, II*, Progr. in Optics, **V**, 1966, p. 333.
[22] KOUYOUMJIAN R. G. *Asymptotic high frequency methods*, Proc. I.E.E.E., **53**, August 1965, p. 864.
[23] KOUYOUMJIAN R. G. and PATHAK P. H. *A uniform geometrical theory of diffraction for an edge in a perfectly conducting surface*, Proc. I.E.E.E., **62**, November 1974, p. 1448.
[24] LUNEBURG R. K. *Mathematical Theory of Optics*, University of California Press, Berkeley, 1966.
[25] MIKHLIN S. G. *Multidimensional Singular Integrals and Integral Equations*, Pergamon, London, 1965.
[26] MULLER C. *Foundations of the Mathematical Theory of Electromagnetic Waves*, Springer, Berlin, 1969.
[27] PAPAS C. H. *Theory of Electromagnetic Wave Propagation*, McGraw-Hill, New York, 1965.
[28] PATHAK P. H. and KOUYOUMJIAN R. G. *An analysis of the radiation from apertures in curved surfaces by the geometrical theory of diffraction*, Proc. I.E.E.E., **62**, November 1974, p. 1438.
[29] RUCK G. T., BARRICK D. E., STUART W. D. and KRICHBAUM C. K. *Radar Cross-section Handbook*, Plenum Press, New York, 1970.
[30] SCHWARTZ L. *Méthodes mathématiques pour les sciences physiques*, Hermann, Paris, 1965.
[31] SCHWARTZ L. *Théorie des distributions* (second edn.), Hermann, Paris, 1966.
[32] SILVER S. *Microwave Antenna Theory and Design*, McGraw-Hill, New York, 1949.
[33] SOMMERFELD A. *Mathematische Theorie der Diffraction*, Math. Ann., **47**, 1896, p. 317.
[34] STRATTON J. A. *Electromagnetic Theory*, McGraw-Hill, New York, 1941.
[35] VAN BLADEL J. *Electromagnetic Fields*, McGraw-Hill, New York, 1964.
[36] VLADIMIROV V. S. *Equations of Mathematical Physics*, Dekker, New York, 1971.
[37] WATSON G. N. *A Treatise on the Theory of Bessel Functions*, Cambridge University Press, 1944.

Second edition

The most important publications, for the purposes of this work, are those concerning above all the geometric theory of diffraction and more specifically the problem of uniform extension of the calculation of the field into singular regions (vicinity of shadow boundaries, caustics, etc). The methods are basically of the UTD (uniform theory of diffraction) type, where the total field is the superposition of the 'geometric' field and of a corrected diffraction field, or of UAT (uniform asymptotic theory) where on the contrary it is the former that is modified.

The following articles may be consulted:

[1] AHLUWALIA D. S. *Uniform asymptotic theory of diffraction by the edge of a three-dimensional body*, S.I.A.M. J. Appl. Math., **18**, 1970, p. 287.

[2] AHLUWALIA D. S., LEWIS R. M. and BOERSMA J. *Uniform asymptotic theory of diffraction by a plane screen*, S.I.A. M. J. Appl. Math., **16**, 1968, p. 783.

[3] BOERSMA J. and RAHMAT-SAMII Y. *Comparison of two leading uniform theories of edge diffraction with the exact uniform asymptotic solution*, Radio Science, **15**, no. 6, 1980, p. 1179.

[4] CIARKOWSKI A., BOERSMA J. and MITTRA R. *Plane wave diffraction by a wedge: a spectral domain approach*, I.E.E.E. Trans. AP., **AP-32**, January 1984, p. 20.

[5] CORNBLEET S. *Geometrical optics reviewed: a new light on an old subject*, Proc. I.E.E.E., **71**, April 1983, p. 471.

[6] DESCHAMPS G. A. *Uniform theories of diffraction by edges. Theoretical methods for determining the interaction of electromagnetic waves with structures*, NATO Advanced Study Institutes series, series **E**: applied sciences, no. 40, 1981, p. 477.

[7] DESCHAMPS G. A., BOERSMA J. and LEE S. W. *Three-dimensional half-plane diffraction: exact solution and testing of theories*, I.E.E.E. Trans. A.P., **AP-32**, March 1984, p. 264.

[8] LEE S. W. and DESCHAMPS G. A. *A uniform asymptotic theory of electromagnetic diffraction by a curved wedge*, I.E.E.E. Trans. A.P., **AP-24**, January 1976, p. 25.

[9] PATHAK P. H., BURNSIDE W. D. and MARHEFKA R. J. *A uniform G.T.D. analysis of the diffraction of electromagnetic waves by a smooth convex surface*, I.E.E.E. Trans. A.P., **AP-28**, September 1980, p. 631.

[10] PATHAK P. H., WANG N., BURNSIDE W. D. and KOUYOUMJIAN R. G. *Uniform G.T.D. solution for the radiation from sources on a smooth convex surface*, I.E.E.E. Trans. A.P., **AP-29**, July 1981, p. 600.

[11] RAHMAT-SAMII Y. and MITTRA R. *A spectral domain interpretation of high-frequency diffraction phenomena*, I.E.E.E. Trans. A.P., **AP-25**, 1977, p. 676.

Part 2

General properties of antennas

Many communication systems use the propagation of electromagnetic waves in natural media. For such systems, antennas constitute indispensable components, the role of which is to ensure the coupling of an electronic circuit and the propagation medium. Part 2 is concerned with the general properties of antennas and considers their characteristics relative to the propagation medium in which they radiate and to the circuit to which they are connected.

The radio link between two antennas depends largely on the propagation medium. Unless otherwise stated, it will be assumed that this medium is homogeneous, isotropic and non-dissipative (free space). In addition, for a long-distance link, it is possible to study the antenna involved in transmission separately from that involved in reception. This separation is useful for study purposes, but does not reflect the existence of two distinct types of antenna. The same antenna is sometimes used simultaneously for transmission and reception.

Chapter 4 is concerned with transmission antennas. The discussion of results from Part 1 allows a description of the radiated field that corresponds well to practical needs. The behaviour of the antenna relative to the circuit that supplies it is considered very briefly, and the example of elementary sources clearly illustrates certain basic aspects of radiation.

In Chapter 5, the discussion of relationships existing between the field and the sources for simple structures—planar or linear—allows the effects of different parameters on an antenna to be explained.

The synthesis of antennas is summarized in Chapter 6.

Chapter 7 concerns reception mode, the properties of which are deduced entirely from the transmission mode, by simple application of the theory of reciprocity. The establishment of an equivalent scheme allows the functioning of any antenna illuminated by an arbitrary electromagnetic field to be analysed in a very general way. The approximations made in dealing with the coupling phenomena are also explained. These phenomena arise notably when the transmission and reception antennas are close to each other. This chapter ends with a study of antenna behaviour in the presence of noise.

All antenna properties are deduced from equivalent sources or currents, both in transmission and reception mode. Their strict determination takes place through the solution of a difficult problem involving boundary conditions.

For certain structures with simple forms, the calculation can be made analytically, with frequent use of special functions, the ease of handling of which however, cannot, always be guaranteed. The development in computer power—in terms both of capacity and rapidity—has tended to favour the use of purely numerical methods, applicable under conditions that are usually unrestrictive to very diverse structures and excitation fields. The aim of Chapter 8 is to give an overview of these methods, both in connection with the formation of equations and the numerical techniques that can be used for their solution.

Chapter 9 concerns experimental techniques, which here refers to a fundamental method of investigation in antenna technique. On the other hand, it gives direct access to values of practical interest; on the other, it constitutes a vital contribution to the numerical approach by suggesting, *a priori*, approximations likely to simplify calculation or by controlling, *a posteriori*, the validity of numerical results.

Chapter 4

Transmitting antennas

A: Introduction to transmission

In transmission an antenna acts as a structure able to assume varied configurations, and supplied by one or more generators. As in Part 1, a distribution of equivalent currents will be assumed to be known on a suitable surface, creating outside the surface a field similar to that of the antenna. The formulae for radiated fields have already been given. The calculation will not be re-explained, but they are used systematically. It is clear that the properties of the field can vary considerably from one antenna to another, or for a given antenna, depending on the point of observation.

Nonetheless, in the far field, the structure of the field can be specified as a locally plane wave, independently of the antenna considered.

Conversely, in the area immediately around the antenna the quasi-static field predominates, and there is a large, concentrated reactive power. This zone is known as the near field or the reactive field. Between these two extreme zones, characterized by very different behaviour, there is a transition zone, which is not clearly defined and the extension of which depends greatly on the structure considered and the operating frequency.

Elementary sources, by virtue of their simplicity, constitute an excellent example in which it is possible to follow the change in field from the near to the far field. Local knowledge of the different components of the electromagnetic field can be advantageously replaced, in many applications, by that of the overall magnitudes, such as gain or directivity, which can be measured directly and characterize the localization of electromagnetic radiation in space. The supply to the structure through a generator can be more or less localized. At low frequencies, therefore, for wire structures, the 'input terminals' of the antenna can be determined without ambiguity. In relation to this input, it is possible to define an input impedance. At high frequencies, the supply is often achieved through a guiding structure. It is then better to characterize the antenna by defining a reflection coefficient. Generally speaking, the behaviour of the antenna relative to the supply circuit constitutes a difficult problem. It will be shown that, in certain simple cases, it can be deduced from the approximate equivalent currents. A general treatment of the problem is not possible until Chapter 8, which concentrates on numerical methods.

B: Far-field radiation

1: General properties

Consider an antenna functioning in transmission mode in a homogeneous medium with permittivity ε and permeability μ. $\vec{J}$ and $\vec{M}$ designate equivalent currents distributed over a suitable closed surface S_a. At any point outside S_a, the field can be calculated with the following relationship (see Part 1, page 30):

$$\vec{E} = \frac{i}{\omega\varepsilon}[\text{grad div} + k^2](\vec{J} * G) + \text{curl}\,(\vec{M} * G) \tag{1}$$

$$\vec{H} = \frac{i}{\omega\mu}[\text{grad div} + k^2](\vec{M} * G) - \text{curl}\,(\vec{J} * G) \tag{2}$$

with $k = \omega\sqrt{\varepsilon\mu}$, the propagation constant, and:

$$G(\vec{x}, \vec{x}') = -\frac{e^{-ik|\vec{x} - \vec{x}'|}}{4\pi|\vec{x} - \vec{x}'|} \tag{3}$$

Considerable simplification can be achieved with the two hypotheses $|\vec{x}| \gg \max|\vec{x}'|$ and $|\vec{x}| \gg \lambda$ (Part 1, page 37). These two conditions allow the far-field zone to be defined. In this zone the field at a distance r in direction $\vec{u}$ is:

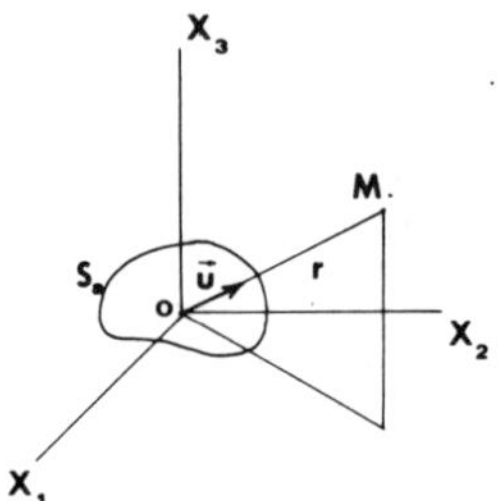

Fig. 1.

$$\vec{E} = \frac{e^{-ikr}}{r}\vec{F}(\vec{u}) + O\left(\frac{1}{r}\right) \tag{4}$$

$$\vec{H} = \frac{1}{\eta}\vec{u} \wedge \vec{E} \tag{5}$$

where $\vec{F}(\vec{u})$ designates the vector radiation characteristic and $\eta = \sqrt{\mu/\varepsilon}$ is the wave impedance. $\vec{F}(\vec{u})$ is related to the currents by the equation:

$$\vec{F}(\vec{u}) = \frac{ik}{4\pi}\int_{S_a} [(\eta\vec{J} \wedge \vec{u}) \wedge \vec{u} - \vec{M} \wedge \vec{u}]\, e^{ik\vec{x}'\cdot\vec{u}}\, dx' \tag{6}$$

In the far field the following important properties arise:

- In the field formulae, there is separation of the angular variables ($\vec{u}$) and the distance r.

• Whatever the distance, the angular distribution of the field is completely determined using only the radiation characteristic.

• Whatever the direction of observation, the field decreases as r^{-1}.

• The set of vectors $\vec{E}$, $\vec{H}$, $\vec{u}$ are mutually orthogonal and right-handed. It follows that the field is transverse. The relationship of one of the components of $\vec{E}$ to that of $\vec{H}$, orthogonal to it, is equal to the wave impedance of the propagation medium, or $\eta = \sqrt{\mu/\varepsilon}$. In vacuo, $\eta = 120\pi \simeq 377\Omega$. The far field is said to comprise a locally plane wave structure. The difference from a 'real' plane wave resides in the attentuation term in r^{-1}, and in the fact that the equiphase surfaces are not plane.

• In the general case $\vec{F}$ is a complex vector with two transverse components. The field is, therefore, elliptically polarized. The polarization plane is perpendicular to the direction of propagation.

• With the above definition adopted for $\vec{F}$, each of its components is measured in volts.

A radiation characteristic could also be defined starting with the magnetic field:

$$\vec{H} = \frac{e^{-jkr}}{r}\,\vec{G}(\vec{u}) + O\left(\frac{1}{r}\right) \tag{7}$$

with

$$\vec{G}(\vec{u}) = \frac{ik}{4\pi}\int_{S_a}\left[\vec{J} \wedge \vec{u} + \frac{1}{\eta}\,(\vec{M} \wedge \vec{u}) \wedge \vec{u}\right] e^{ik\vec{x}'\cdot\vec{u}}\,dx' \tag{8}$$

The components of $\vec{G}$ are measured in amps.

The following relationship exists between $\vec{F}$ and $\vec{G}$:

$$\vec{F}(\vec{u}) = \eta\vec{G}(\vec{u}) \wedge \vec{u} \tag{9}$$

Note also that the norms associated with $\vec{F}$ and $\vec{G}$ are such that:

$$F' = \eta G' \tag{10}$$

• It is sometimes helpful to use a reduced dimensionless characteristic by normalizing the distribution of $\vec{J}$ and $\vec{M}$ relative to the value of one of the components of $\vec{J}$ or $\vec{M}$ at a particular point on the antenna surface. The normalization of $\vec{J}$—or of $\vec{M}$—automatically brings about that of $\vec{M}$—or of $\vec{J}$—since $\vec{J}$ and $\vec{M}$ are not independent on a closed surface.

2: Translation theorem

When an antenna is displaced by translation from point O_1 to point O_2, the field $\vec{E}_2$ radiated in the new position is expressed quite simply as a function of $\vec{E}_1$, the field radiated in the initial position. In effect, r_i and $\vec{u}_i$ can be considered as the distance and direction of the observation point located relative to O_i and $\vec{\delta} = \overrightarrow{O_1O_2}$, as the translation vector (Figure 2). As a result of the translation, and taking account of the particular dependence of the field on the distance, it follows that:

$$\vec{E}_1 = \frac{e^{-ikr_1}}{r_1} \vec{F}(\vec{u}_1) + O\left(\frac{1}{r_1}\right)$$
$$\vec{E}_2 = \frac{e^{-ikr_2}}{r_2} \vec{F}(\vec{u}_2) + O\left(\frac{1}{r_2}\right) \tag{11}$$

If the observation point is situated at a great distance, then to first order: $\vec{u}^1 = \vec{u}_2 = \vec{u}$ and $r_2 = r_1 - \vec{\delta} \cdot \vec{u}$.

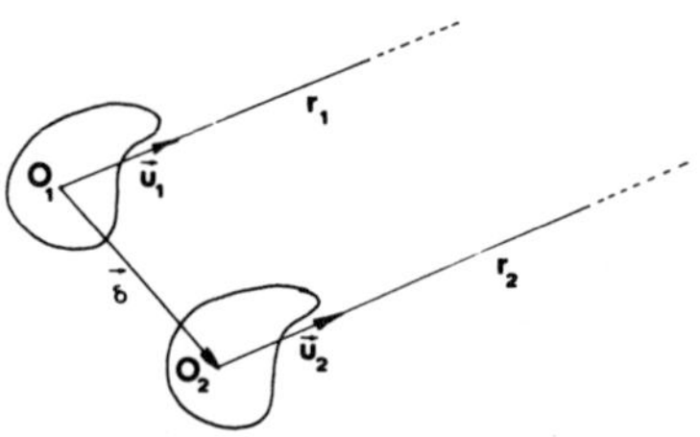

Fig. 2. *Translation of an antenna.*

The approximation $r_1 = r_2$ is admissible in the attenuation term provided that δ is small compared to r_1 and r_2. It will also be admissible in the phase term if δ is small compared to λ, but this is too restrictive.

It follows, therefore, that:

$$\vec{E}_2 = \vec{E}_1 \, e^{ik\vec{\delta} \cdot \vec{u}} \tag{12}$$

Hence the theorem:

When an antenna is translated, the radiated field undergoes a *shift* measured by the projection of the displacement along the direction of observation, while the modulus remains unchanged.

For a given antenna, therefore, the choice of origin is somewhat arbitrary.

3: Superposition theorem

The field is deduced from the currents by means of a linear operation. This property is true everywhere, and particularly in the far field. The theorem of superposition applied to the phenomena of radiation can therefore be expressed as follows:

If $(\vec{E}_k, \vec{H}_k)$ designates the field created by the independent sources $(\vec{J}_k, \vec{M}_k)$, the source $(\vec{J} = \Sigma\vec{J}_k, \vec{M} = \Sigma\vec{M}_k)$ creates the field $(\vec{E} = \Sigma\vec{E}_k, \vec{H} = \Sigma\vec{H}_k)$.

The theory is applied without qualification for an antenna or set of antennas supplied by several generators. $(\vec{E}_k, \vec{M}_k)$ therefore represents the partial field resulting from the connection of the kth generator, with the other generators being replaced by an impedance equal to their internal impedance.

Strictly speaking, the theory cannot be applied when the configuration of each of the individual cases is not identical to that of the problem as a whole. For example, the field created by a set of antennas is not the sum of the fields created by each of the antennas considered in isolation. Quite often, however, as a preliminary approximation it will be assumed that the theorem of superposition can be applied to the above problem, which is equivalent to ignoring the coupling which arises between the various antennas (weak coupling hypothesis).

4: Representations of the vector radiation characteristic

The complete determination of the radiated field in a space implies that, for all directions $\vec{u}$, the transverse components of $\vec{F}$ are known with their amplitude and phase. In fact, to form a clear idea of the distribution of the field, it is often sufficient to consider the norm F' or its square F'^2, the connection of which with the radiated power will be explained.

A single characteristic radiation surface can be constructed, in a system of spherical or rectangular coordinates (Figure 3).

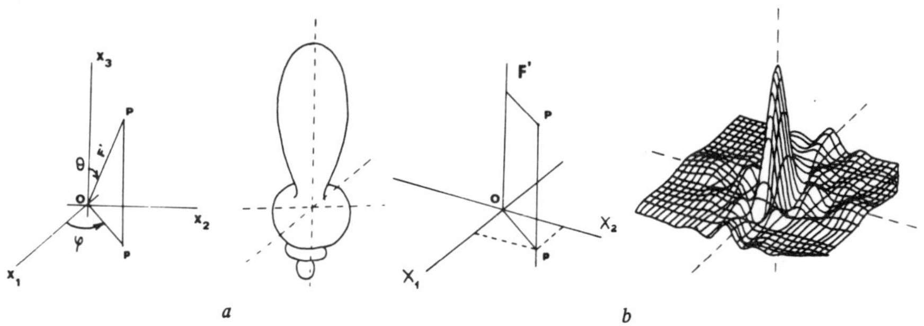

Fig. 3. *Representations of the vector radiation characteristic* $F'^2 = \vec{F} \cdot \vec{F}^*$.

The choice of one or the other system depends on the location of the radiated field in space. If the radiated field is distributed over the whole space (weakly directive antenna), the spherical system is the best choice. The rectangular system is more suitable when the field is concentrated around a particular direction (directive antenna). The values of F' are plotted on linear, quadratic or logarithmic scales.

Generally, the construction of this surface is awkward. It is usually preferable, therefore, to form sections or projections. The curves thus obtained are called radiation diagrams. The sections are taken in the planes of symmetry, if there are any, or else, in a more conventional way, in the orthogonal planes corresponding to the planes of field polarization in a preferred direction—for example, *E*-plane, *H*-plane. The polar representation is very evocative. Cartesian representation is useful for the comparison of diagrams: with logarithmic scales, any readjustment of a diagram due to a multiplicative factor will require only one vertical translation.

Projections are useful for highly directive antennas. They are carried out on a plane normal to the preferred direction of radiation. The radiation diagram appears in the form of level curves denoting the antenna response, which can be interpreted immediately. For directive antennas, the diagrams show a peak, called the principal lobe. The other peaks, generally much lower than the first, are called sidelobes. The angular width of the principal lobe is defined at a certain attenuation: the term used is, for example, 3 dB or 10 dB beamwidth. The importance of the sidelobes is measured by the highest sidelobe level, which generally measures, in dB, the level of the highest sidelobe relative to that of the principal lobe.

C: Radiated power

1: Active and reactive power

The problem of radiated power has already been discussed in Part 1 (page 14). The results, which will allow us to write the balance of power, will be rapidly summarized here. Let Ω be a domain enclosing the antenna, with boundary Σ. The complex power supplied to the antenna is denoted by W_a, the active power dissipated in Ω is denoted by W'_d. The power radiated across Σ is:

$$W_r = W'_r + iW''_r = \frac{1}{2}\int_{\Sigma} (\vec{E} \wedge \vec{H}^*)\cdot\vec{n}d\sigma \tag{13}$$

where $\vec{n}$ is a unit vector directed along the external normal. The application of Poynting's theorem at Ω gives:

$$W_a = W_r + W'_d + 2i\omega(T_H - T_E)_{\Omega} \tag{14}$$

where $(T_H - T_E)_{\Omega}$ designates the average magnetic excess within Ω:

$$(T_H - T_E)_{\Omega} = \frac{1}{4}\int_{\Omega} (\mu H'^2 - \varepsilon E'^2)\, dv \tag{15}$$

By taking the real part of (14), $W'_a = W'_r + W'_d$.

The active power W'_r radiated across Σ is equal to the difference between the active power provided by the generator and the power that is dissipated. If the propagation medium is lossless, the power is dissipated in the antenna itself and therefore independently of Σ. It follows that the radiated active power is also independent of Σ.

This result is important: the surface Σ can, in fact, be chosen to allow the calculation of W'_r to be carried out in the simplest way. This is the case for a sphere S, with large radius, centred on the antenna, for which the far-field expressions can be used. It follows that:

$$W'_r = \frac{1}{2\eta}\int_{S_x} E'^2\, dS = \frac{1}{2\eta}\int_{4\pi} F'^2(\vec{u})\, d\bar{\omega}(\vec{u}) \tag{16}$$

where $d\bar{\omega}$ denotes the solid angle element in the direction $\vec{u}$. But the calculation is still possible on the surface of the antenna itself, with the equivalent currents:

$$W'_r = \frac{1}{2}\,\mathrm{Re}\left(\int_{S_a} \vec{J}\cdot\vec{M}^*\, dS\right) \tag{17}$$

By taking the imaginary part of (14):

$$W''_a = W''_r(\Omega) + 2\omega(T_H - T_E)_\Omega \tag{18}$$

The application of Poynting's theory in the domain Ω' outside Σ would give:

$$O = -W''_r(\Omega) + 2\omega(T_H - T_E)_{\Omega'} \tag{19}$$

Thus, by taking the sum of (18) and (19):

$$W''_a = 2\omega(T_H - T_E)_{\Omega\cup\Omega'} \tag{20}$$

The average magnetic excess in the whole space is bounded and equal to the reactive power delivered by the generator. The result of this is that the excess relative to Ω' tends towards 0 when Σ tends towards S_∞. The same applies for W''_r because of (19). The result is that the radiated reactive power depends on the surface chosen to calculate it; it decreases farther away from the antenna, as might be predicted from the expressions for radiated fields at great distances. The situation is, therefore, similar to that which arises during the excitation of a wave in a guide. The source excites a large number of modes in its surroundings. Only the modes capable of propagation in the guide transport the active power. Evanescent modes attenuate exponentially and contribute to the localization of reactive power around the source.

2: Gain and directivity

The expression of W'_r as a function of F' (16) suggests the definition of a power density by unit solid angle $\sigma(\vec{u})$, such that:

$$\sigma(\vec{u}) = \frac{dW'_r(\vec{u})}{d\bar{\omega}(\vec{u})} = \frac{E'^2 r^2}{2\eta} = \frac{F'^2}{2\eta} \tag{21}$$

This density has no concrete physical significance since Poynting's theorem makes sense in its integral form only for a closed surface. The efficiency of radiation in direction $\vec{u}$ can be measured by the relationship:

$$D(\vec{u}) = \frac{\sigma(\vec{u})}{\sigma_m} \tag{22}$$

where σ_m denotes the average power density corresponding to perfectly isotropic radiation of the radiated power W'_r:

$$\sigma_m = \frac{W'_r}{4\pi} \tag{23}$$

The ratio $D(\vec{u})$ is called the directivity of the antenna in direction $\vec{u}$:

$$D(\vec{u}) = \frac{4\pi\sigma(\vec{u})}{W'_r} = 4\pi \frac{F'^2(\vec{u})}{\displaystyle\int_{4\pi} F'^2(u)\, d\bar{\omega}(\vec{u})} \tag{24}$$

It is sometimes preferable to evaluate the efficiency of radiation in terms of the effective power provided by the source. The gain $G(\vec{u})$ is therefore defined by:

$$G(\vec{u}) = \frac{4\pi\sigma(\vec{u})}{W'_a} = \rho D(\vec{u}) \tag{25}$$

where ρ denotes the efficiency loss.

$$\rho = \frac{W'_r}{W'_a} = \frac{W'_r}{W'_r + W'_d} \tag{26}$$

For an antenna without losses, the gain is identical to the directivity. This is often the case at high frequencies.

A number of points should be made:

- The directivity allows the localization of the radiated power to be characterized. High directivity signifies that the radiation is, for the most part, localized in a restricted solid angle. As will be seen later, directivity is essentially linked to the ratio of the antenna dimension to the wavelength. Large antennas are generally highly directive and, conversely, small antennas are non-directive.
- Gain results from a directive antenna by more economic use of the power supplied by the source. The gain in a given direction $\vec{u}$ could also be defined by the ratio of the power that would have to be supplied to the antenna to that which would have to be supplied to an isotropic antenna to obtain the same radiated power in that direction.
- The definitions of gain and directivity refer to the isotropic antenna. Taking account of the vector nature of the electromagnetic field, an antenna of this type that would radiate isotropically over all space is obviously a physical impossibility. Since in practical terms it may be inconvenient to express the gain of an antenna with respect to a physical impossibility, it is sometimes more convenient to compare an antenna with another, realizable, reference antenna, which must then be defined. The directivity of antenna 1 would thus be defined in relation to antenna 2, by:

$$D_{12}(\vec{u}) = \frac{D_1(\vec{u})}{D_2(\vec{u})} \tag{27}$$

The choice of reference antenna depends on the frequency considered. Its structure is generally simple, so that its directivity can be calculated with a high level of precision. At low frequencies, the short dipole or half-wave dipole is often used. At microwave frequencies, horns are more common.

• Gain and directivity are dimensionless numbers, defined by power ratios. These values are often expressed in decibels. Fo directivity, therefore:

$$D_{\mathrm{dB}}(\vec{u}) = 10 \log_{10} D(\vec{u}) \tag{28}$$

Note

If the gain $G(u)$ of an antenna and the power supplied W'_a are known, the norm of the radiated field can be calculated at distance r and in direction $\vec{u}$:

$$E'(r, \vec{u}) = \sqrt{\frac{\eta}{2\pi}} \frac{\sqrt{G(\vec{u}) W'_a}}{r} \tag{29}$$

3: Input impedance—radiation resistance

(a) General definitions

This section is concerned with characterizing the charge brought by the antenna to the excitation circuit. The simplest case is that of antennas for which two input terminals can be defined without ambiguity, such that in transmission mode there is a voltage V_e between them when a current I_e is flowing. The antenna can be represented by a dipole—in the sense used in circuit theory—the input impedance of which would be:

$$Z_e = \frac{V_e}{I_e} = R_e + iX_e \tag{30}$$

More generally, an impedance Z and a current I can be associated with the total powr W_a, supplied by the generator, such that:

$$W_a = \frac{1}{2} Z I'^2 \tag{31}$$

If $I = I_e$, Z is identical with the input impedance Z_e. This energy definition provides a method for calculating Z_e. Its real part is defined without ambiguity:

$$W'_a = W'_r + W'_d = \frac{1}{2}(R_r + R_d) I'^2_e = \frac{1}{2} R_e I'^2_e \tag{32}$$

R_r and R_d correspond, respectively, to the radiated active power and to the power dissipated in the antenna. The radiation resistance R_r is deduced simply from the vector radiation characteristic:

$$R_r = \frac{1}{I'^2_e} \int_{4\pi} F'^2(\vec{u})\, d\varpi(\vec{u}) \tag{33}$$

The ratio F'/I'_e can be interpreted as a reduced characteristic calculated from the normalized equivalent sources in relation to the electric current—and therefore to the density of electric current—at the antenna input terminals. It was shown that the normalization of $\vec{J}$ automatically brings about that of $\vec{M}$, since on a closed surface the distributions of $\vec{J}$ and $\vec{M}$ are not independent. The calculation of R_r therefore uses values from the far field, which are known to be obtained themselves from the approximate distributions of current on the antenna. Very generally, a radiation resistance can be defined from (31) for an arbitrary current. If $I \neq I_e$, the radiation resistance is no longer the input resistance. The loss resistance R_d is deduced from the known field in the antenna:

$$R_d = \frac{1}{I_e'^2} \int_{\Omega_a} \vec{E} \cdot \vec{J}^* \, dv \tag{34}$$

For antennas with low losses, a perturbation method can be used. In all cases, the normalization of the field relative to the input current is useful. It allows, in particularly, the efficiency to be expressed in the form of a resistance ratio:

$$\rho = \frac{R_r}{R_r + R_d} \tag{35}$$

For the imaginary part of the input impedance, the problem is less simple. Evaluation of the flux of Poynting's vector across a surface Σ surrounding the antenna gives:

$$W_a'' = W_r''(\Sigma) + 2\omega(T_H - T_E)_\Omega = \frac{1}{2} X_e I'^2 \tag{36}$$

In this case, any splitting of X_e would be purely arbitrary since it would rest on the choice of a surface Σ.

So far, all reasoning has been based on the impedances, but there is no reason why the conductances should not be used. It is the structure of the antenna that justifies the choice of one representation or the other.

An antenna is said to be resonant when the input reactance—or the conductance—cancels itself out. The intensity of resonance can be evaluated with the generalized definition of the Q factor. So for a series resonance:

$$Q = \frac{1}{2R_{eo}} \frac{\partial X_e}{\partial \omega}\bigg|_{\omega_0} \tag{37}$$

where ω_0 corresponds to the resonance frequency and R_{eo} designates the input resistance at resonance.

At microwave frequencies, the idea of voltage is less applicable. The antennas are, however, generally fed by waveguide structures whose circuit properties are less localized. In the same way as for usual circuits, it is helpful to consider the antenna as constituting a discontinuity, characterized by a reflection coefficient—or several, in the case of multimode structures—referred to an arbitrarily chosen plane of reference. Similar considerations to those explained above allow the reflection coefficient to be calculated from the radiated power.

(b) Notes on methods of calculation

The calculation of input impedance is generally difficult. It is based on a good description of the fields surrounding the antenna. Two classes of calculation method can be distinguished: the first requires the exact currents to be known, the second only the approximate currents.

In the first case, the currents can be calculated analytically for relatively simple structures (linear or conical antennas), or numerically for far less restrictive conditions (see Chapter 8). For this type of approach, the current is determined over the whole antenna, and particularly at the input terminals. The input impedance is deduced from equation (30).

In the second case, the input impedance is deduced indirectly from an approximate distribution of the currents. Taking into account the energy definition of the input impedance, for an antenna without losses:

$$Z_e = \frac{1}{I_e'^2} \int_{S_a} \vec{J}^* \cdot \vec{M} \, dS \tag{38}$$

It is possible to calculate the electric field—and therefore $\vec{M}$—created on S_a by a given distribution $\vec{J}$. The above expression provides a means of calculating Z_e. This process is known as the induced electromotive force method. Note also that there are variational methods that give rise to expressions of the input impedance that are not affected by first-order errors in the currents. These will not be further discussed here for two reasons: first, they require a great deal of development and second, they are somewhat less used since the advent of numerical methods. Interested readers are referred to Collin and Zucker, and to Galejs.

(c) Correlative antennas

Consider two correlative current distributions $\vec{J}$ and $\vec{M}$ such that $\vec{J} = \eta \vec{M}$ at any point on the antenna surface. Such distributions can be respectively envisaged and realized with electric currents circulating in a thin metal plate and magnetic currents existing in a slot of the same shape in a conductive

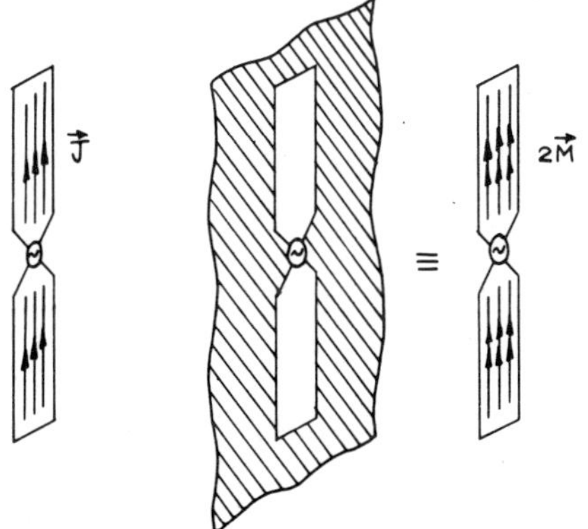

Fig. 4. *Correlative antennas.*

plane (Figure 4). It is known that the slot and the conductive plane together are equivalent to the distribution $2\vec{M}$ *in vacuo* (see Part 1, page 57).

The power W_{re} radiated by the currents $\vec{J}$ can be expressed as:

$$W_{re} = \frac{1}{2\eta} F_e'^2 I_0^2 = \frac{1}{2} R_r I_0^2 \tag{39}$$

where R_r designates the radiation resistance related to the current I_0, and $\vec{F}_e$ is the reduced vector characteristic of radiation:

$$\vec{F}_e(\vec{u}) = \frac{ik}{4\pi} \int_{S_a} \eta \left(\frac{\vec{J}}{I_0} \wedge \vec{u} \right) \wedge \vec{u}\, e^{ik\vec{r}' \cdot \vec{u}}\, dr' \tag{40}$$

For the magnetic distribution, the radiated power is:

$$W_{rm} = \frac{1}{2\eta} F_m'^2 V_0^2 = \frac{1}{2} G_r V_0^2 \tag{41}$$

where G_r denotes the radiation conductance related to the voltage V_0 and $\vec{F}_m$ is the reduced characteristic:

$$\vec{F}_m(\vec{u}) = -\frac{ik}{4\pi} \int_{S_a} \left(\frac{2\vec{M}}{V_0} \wedge \vec{u} \right) e^{ik\vec{r}' \cdot \vec{u}}\, dr' \tag{42}$$

From this it can be deduced:

$$\frac{R_r}{G_r} = \frac{F_e'^2}{F_m'^2} \tag{43}$$

If V_0 and I_0 correspond to the same point on the antenna, then $V_0 = \eta I_0$. Therefore, the following relationship between the radiation resistance and conductance holds:

$$\frac{R_r}{G_r} = \frac{\eta^2}{4} \tag{44}$$

(d) Symmetrical antennas

In practice, the image principle is very often used to replace a symmetrical antenna by a half-antenna in the presence of a conductive plane (Figure 5). The half-antennas and the plane are equivalent to a complete antenna fed by twice the magnetic surface current (Part 1, page 57). For the same input current, therefore, the voltage applied for the whole antenna is double the voltage

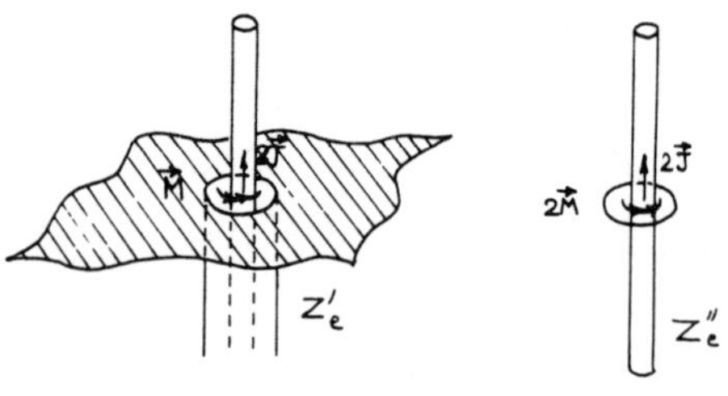

Fig. 5.

applied for the half-antenna. If the input impedances for the half and whole antennas respectively are called Z'_e and Z''_e:

$$Z'_e = \frac{Z''_e}{2} \tag{45}$$

D: The elementary source case

Elementary sources were defined in Chapter 1, page 31. They correspond to current distributions of the type $\vec{K} = \vec{K}_0\,\delta_x$. The study of the field radiated by these sources is of great interest. They constitute a simple teaching example on which the main properties of the field radiated by the antenna can be illustrated and in addition, because of their simplicity, the field can be calculated at any point in space without approximation. The sources considered here are not hypothetical, but can be realized in practice, to a high degree of approximation, using simple structures. Finally, they constitute a basis for the analysis of any distribution of currents: the characteristics of many antennas can be deduced almost instantly.

Only one electric source (a doublet) will be considered here, characterized by the density of current $\vec{J}_0$ situated at the origin of a coordinate system $Oxyz$, and directed along the axis Oz. It is easy to prove that J_0 has the dimensions of the product of a current and a length. Let $J_0 = I_0 l$. The calculation of components in the radiated field at a point defined by its spherical coordinates r, θ, φ has already been carried out (see Chapter 1, page 34). Taking account of the symmetry, the field admits only three components:

$$\begin{aligned}
E_r &= \frac{i\omega\mu}{4\pi}\left(\frac{2i}{kr} + \frac{2}{k^2r^2}\right) J_0 \cos\theta\,\frac{e^{-ikr}}{r} \\
E_\theta &= -\frac{i\omega\mu}{4\pi}\left(-1 + \frac{i}{kr} + \frac{1}{k^2r^2}\right) J_0 \sin\theta\,\frac{e^{-ikr}}{r} \\
H_\varphi &= +\frac{ik}{4\pi}\left(1 - \frac{i}{kr}\right) J_0 \sin\theta\,\frac{e^{-ikr}}{r}
\end{aligned} \tag{46}$$

The far field is obtained from the requirement $r \gg \lambda$, or $kr \gg 1$, the condition $r \gg r'_{\max}$ having no meaning since the antenna is assumed to be a point-source. It follows that:

$$\begin{aligned}
E_r &= O\left(\frac{1}{r^2}\right) \\
E_\theta &= \frac{i\omega\mu}{4\pi} J_0 \sin\theta\,\frac{e^{-ikr}}{r} + O\left(\frac{1}{r}\right) \\
H_\varphi &= \frac{i\omega\varepsilon\eta}{4\pi} J_0 \sin\theta\,\frac{e^{-ikr}}{r} \to O\left(\frac{1}{r}\right)
\end{aligned} \tag{47}$$

The locally plane wave nature is immediately verified. The radial component is negligible compared to the transverse components. The vector radiation characteristic is carried by the unit vector $\vec{e}_0$:

$$\vec{F}(\theta) = \frac{ik\eta}{4\pi} I_0 l \sin\theta \vec{e}_\theta \tag{48}$$

The field is therefore linearly polarized. The three-dimensional radiation pattern is a torus with axis Oz. The radiation is maximal in the plane xOy and cancels in the far field along the z-axis. Surfaces of constant phase are spheres centred on the origin.

Around the origin, the field components vary as their highest negative power of r. It is helpful to introduce the bipolar moment $P = -J_0/i\omega$ (Chapter 1, page 31) of the doublet that appears quite naturally in the equation of current continuity. It follows that:

$$\begin{aligned} E_r &= -\frac{1}{2\pi\varepsilon}\frac{P\cos\theta}{r^3} + O\left(\frac{1}{r^2}\right) \\ E_\theta &= -\frac{1}{4\pi\varepsilon}\frac{P\sin\theta}{r^3} + O\left(\frac{1}{r^2}\right) \\ H_\varphi &= \frac{1}{4\pi}\frac{J_0\sin\theta}{r^2} + O\left(\frac{1}{r}\right) \end{aligned} \tag{49}$$

For the electric field, there is an electrostatic-type distribution, independent of frequency, decreasing as r^3 and with rectilinear polarization. For the magnetic field, a contribution is recognizable from a current element $J_0 = I_0 l$, conforming to Biot and Savart's law, decreasing as r^2. At any point close to the source, the approximation of the quasi-stationary states is verified.

Between the two extreme zones where the behaviour of the field can be described very simply, there is a transition zone in which the field structure is relatively more complicated. In particular, the electric field is polarized elliptically. It will be noted, however, that at any distance there is a separation of variables r and θ. This situation is not the same for all antennas.

At distance r, the radial component of Poynting's vector is:

$$P_r = \frac{\eta k^2 \sin^2\theta I_0^2 l^2}{2(4\pi r)^2}\left(1 - \frac{i}{k^3 r^3}\right) \tag{50}$$

The power radiated across a sphere of radius r is deduced as follows:

$$W_r = W'_r + iW''_r = \frac{\eta k^2 I_0^2 l^2}{12\pi}\left(1 - \frac{i}{k^3 r^3}\right) \tag{51}$$

It can be proved that the active power is independent of r. The reactive power is localized around the origin. The negative sign indicates that the power is electric in type (negative magnetic excess). To fix the orders of magnitude, the active power is equal to the reactive power for $kr = 1$, so $r = \lambda/2\pi$. The radiation resistance, related to the current I_0 which here plays the role of input current, is, *in vacuo*:

$$R_r = \frac{2\pi}{3}\eta\left(\frac{l}{\lambda}\right)^2 \quad (52)$$

This formula is valid only for $1 \ll \lambda$. As an indication, $R_r = 0.08\Omega$ for $l = \lambda/100$. It can be imagined that, in these conditions, the matching of the generator to the antenna presents serious problems, and that the losses of the matching elements, however small, rapidly become very important. The input reactance is capacitive and in the limit $r \to 0$ becomes infinite. Directivity is maximal in plane xOy where it reaches 1.5 or 1.8 dB. By duality, results relating to the elementary magnetic source are also obtained, and this can be realized in practice using a loop of radius $a \ll \lambda$, through which a constant current I_0 flows.

Bibliography to Chapter 4

ANGOT. *Compléments de mathématiques*, Editions de la Revue d'Optique, 1959.

COLLIN R. E., ZUCKER F. *Antenna Theory*, McGraw-Hill, 1969.

DUBOST G., ZISLER S. *Antennes à large bande*, Masson 1976.

GALEJS J. *Antennas in Inhomogeneous Media*, Pergamon Press, 1969.

HANSEN R. C. *Microwave Scanning Antennas*, Academic Press, 1964.

HARRINGTON R. F. *Time-harmonic Electromagnetic Fields*, McGraw-Hill, 1961.

JASIK H. *Antenna Engineering Handbook*, McGraw-Hill, 1961.

JORDAN E. C., BALMAIN K. G. *Electromagnetic Waves and Radiating Systems* (2nd edn.), Prentice-Hall, 1968.

KRAUS J. D. *Antennas*, McGraw-Hill, 1950.

SCHELKUNOFF S. A. *Electromagnetic Waves*, D. Van Nostrand, 1945, Chap. XI: 'Antenna theory', pp. 441–479.

SCHELKUNOFF S. A. *Antenna Theory and Practice*, John Wiley & Sons, 1952.

SILVER S. *Microwave Antenna Theory and Design*, McGraw-Hill, 1948, R.L.S.S. Series.

THOUREL L. *Les antennas*, Dunod, 1971.

VAN BLADEL J. G. *Electromagnetic Field*, McGraw-Hill, 1964.

WALTER C. H. *Traveling-wave Antennas*, McGraw-Hill, 1965.

Chapter 5

Examples of applications

A: Introduction

The results of Sections 2 and 3, preceding, are sufficiently general to apply to all antennas, whatever their structure. If only linear or plane structures, from which a great many common antennas are derived, are considered, other important properties can be established. For such structures, it is helpful to consider the radiating structure as an assembly of independent antennas that have the same characteristics and can be derived from each other by translation.

In this way a thin linear antenna can be broken down into colinear electric doublets, the moments of which are proportional to the current. Another example can be seen in a radiating aperture, which, according to the hypothesis of locally plane wave illumination, is equivalent to a plane distribution of electric and/or magnetic doublets of parallel moments. In both these cases, the antenna can be replaced by continuous distributions of elementary sources.

The discontinuous case is also of great practical importance: it corresponds to antenna arrays formed by a large number of identical parallel antennas. These simple antennas can form linear or plane arrays. The way in which the transition can be made between the continuous and discontinuous cases will be explained below.

B: Linear structures

1: Statement of the problem

A linear distribution of elementary sources, aligned along the axis Oz in the coordinate system $Oxyz$, will be considered (Figure 6). The extent of the sources is limited to the interval $\{-l/2, +l/2\}$. The moments are all oriented parallel to the unit vector $\vec{e}_s$. Only electric sources are considered here. The results of magnetic sources can be deduced by duality.

The density of the sources can, therefore, be represented using the expression:

$$\vec{J}(x, y, z) = J_0(z)\,\delta(x)\,\delta(y)\left[U\left(z+\frac{l}{2}\right) - U\left(z-\frac{l}{2}\right)\right]\vec{e}_s \qquad (1)$$

Despite its apparent simplicity, this model—or its dual— can be applied to many antennas in current use: rectilinear wire structures, slots with negligible transverse dimensions relative to their length and to the wavelength, etc.

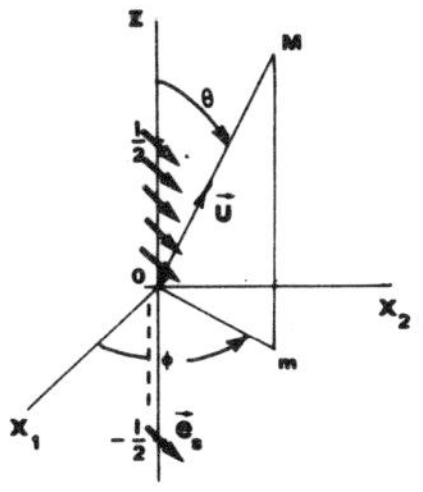

Fig. 6.

2: Basic formula

By direct application of formula (6) from Chapter 4, the vectorail radiation characteristic can be expressed in the form:

$$\vec{F}(\vec{u}) = f(\cos\theta)\vec{F}_e(\vec{u}) \tag{2}$$

with:

$$\vec{F}_e(\vec{u}) = \frac{ik\eta}{4\pi} J_0(o) l(\vec{e}_s \wedge \vec{u}) \wedge \vec{u} \tag{3}$$

$$f(\cos\theta) = \frac{1}{l}\int_{-l/2}^{+l/2} \frac{J_0(z)}{J_0(o)} e^{iks\cos\theta}\, dz \tag{4}$$

This gives the following preliminary result: The vector radiation characteristic of the set of sources appears in the form of the vector radiation characteristic $\vec{F}_e$ of an elementary source located at the origin, multiplied by a scalar function f which depends only on the relative distribution of the intensity of the moments.

For this reason, the function f is often called the alignment function.

It is clear that $\vec{F}_e(\vec{u})$ is identical to the characteristic of a doublet with moment equal to $J_0(o)l$, situated at the origin and oriented along $\vec{e}_s$. In the specific case of $\vec{e}_s = \vec{e}_z$ the expression for the characteristic of an electric dipole defined in equation (48) of Chapter 4 can be used. The function f is independent of the orientation of the moments. It is often helpful to consider it as the radiation characteristic of scalar sources of intensity $J_0(z)$ distributed along axis Oz (acoustic sources, for example). It will be noted that f is independent of φ.

By making changes of variable:

$$\alpha = kl\cos\theta \qquad \xi = \frac{z}{l} \tag{5}$$

the alignment factor can be expressed as follows:

$$g(\alpha) = f\left(\frac{\alpha}{kl}\right) = \int_{-\infty}^{+\infty} \mathcal{J}(\xi)\, e^{i\alpha\xi}\, d\xi \tag{6}$$

where the reduced current distribution $\mathcal{J}(\xi)$ is defined by:

$$\mathcal{J}(\xi) = \begin{cases} \dfrac{J_0\left(\dfrac{\xi l}{2}\right)}{J_0(o)} & |\xi| \leqslant \dfrac{1}{2} \\ 0 & |\xi| > 1/2 \end{cases} \tag{7}$$

The second result is therefore found: The function f appears as the Fourier Transform of the distribution law of the dipole moments. The inverse transform would involve values of f for all the values of α. But α is physically limited to the values between $\pm kl$. The values outside this interval correspond to imaginary angles (see Part 1, page 45).

The existence of a Fourier Transform between f and $\mathcal{J}$ (see Part 1) is fundamental in antenna theory. It allows the treatment, in standard terms, of the influence of current distribution on the radiated field—a problem of analysis—and allows the distribution of current necessary to obtain a given field to be determined—a problem of synthesis (see Chapter 6).

For simply linear structures, many excitations can be envisaged, allowing various distributions of $J(z)$ to be obtained—or correspondingly of $K(z)$—both in terms of amplitude and phase. In the following paragraphs, two important cases will be discussed. The first is that of travelling wave antennas, the second is that of standing wave antennas. These two types are often used in practice. Some examples will allow their particular structures to be described.

3: Travelling wave antennas

(a) Calculation of the alignment function

For such antennas, the distribution $J(z)$ can be interpreted as the result of propagation of a wave with a complex propagation constant. It is, however, the role of the phase constant that is essential. Consider, first of all, the case of a distribution with uniform amplitude of the type:

$$J(z) = J_0\, e^{-i\beta z} \qquad J_0,\ \beta \text{ real} \tag{8}$$

From this an alignment function of the $(\sin x)/x$ type can be deduced immediately:

$$f(\cos\theta) = \operatorname{sinc}(x) = \frac{\sin x}{x} \tag{9}$$

with

$$x = \frac{kl}{2}(\cos\theta - p), \quad p = \frac{\beta}{k}$$

As will be seen in the following, it can be useful to consider f as the radiation characteristic of scalar sources distributed along the axis Oz. Since f has rotational symmetry about axis Oz, the associated characteristic surface is completely defined by its meridian.

Before constructing this meridian, the important properties of the sinc function sin x/x (Figure7) will be summarized. This function has its maximum value 1 at $x = 0$. It presents extrema of decreasing amplitude, alternatively positive and negative. The level of the first of these extrema is 13.3 dB below the maximum. The width of the principal lobe, measured by the distance between the two first zeros, is $\Delta x = 2\pi$. It is double the width of each of the secondary lobes and corresponds more or less to double the width of the principal lobe measured at -3 dB.

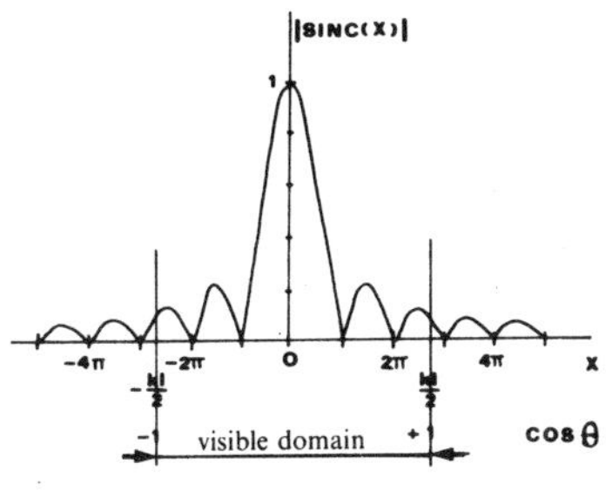

Fig. 7.

The meridian of the characteristic surface associated with f is deduced from a simple graphic construction, as shown in Figure 8. Note that only the interval corresponding to $|\cos\theta| \leqslant 1$ can be represented and is, effectively, observable. For this reason this interval is called the visible domain, as opposed to the invisible domain, which would correspond to $|\cos\theta| > 1$ (complex angles). Directions in which f is cancelled out can be immediately located and the meridian can be constructed point by point without difficulty. It is thus possible to reveal the influence of various factors.

(b) Influence of the length

For $p = 0$, the diagram is symmetric about direction $\theta = \pi/2$. The visible domain corresponds to $|x| \leqslant kl/2$. The extent of this domain is therefore directly proportional to the ratio l/λ. For $l \ll \lambda$, the function f remains practically constant throughout the visible domain. This means that the vector radiation characteristic of the alignment is essentially that of the elementary source, $\vec{F}(\vec{u}) \simeq \vec{F}_e(\vec{u})$. When the ratio l/λ increases, the function f decreases at the limits of the visible domain and is cancelled for $l = \lambda$. If l/λ is further increased, an increasing number of sidelobes appear in the visible domain (Figure 9). When l/λ is very large compared to 1, the diagram is basically centred about the direction $\theta = \pi/2$, so that the total characteristic of the alignment is largely that of the alignment function. Two important results follow from this discussion. The first is that the alignment of sources can be far more directive than the elementary source if $l \gg \lambda$. The alignment function is then almost entirely contained in the visible domain. The second is that for $l \ll \lambda$, a large part of the alignment function is located in the visible domain.

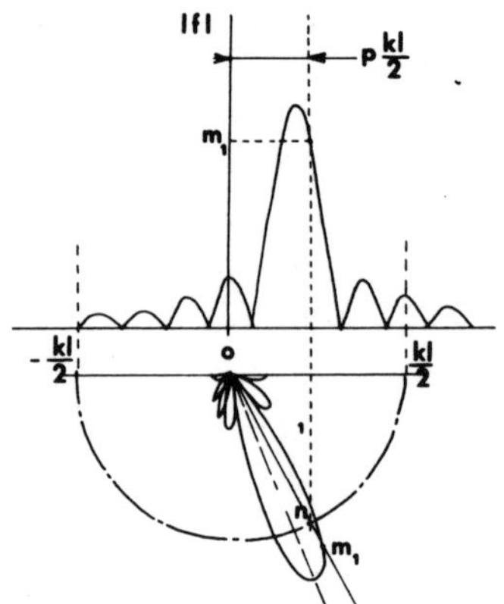

Fig. 8. *Graphic construction of the radiation characteristic.*

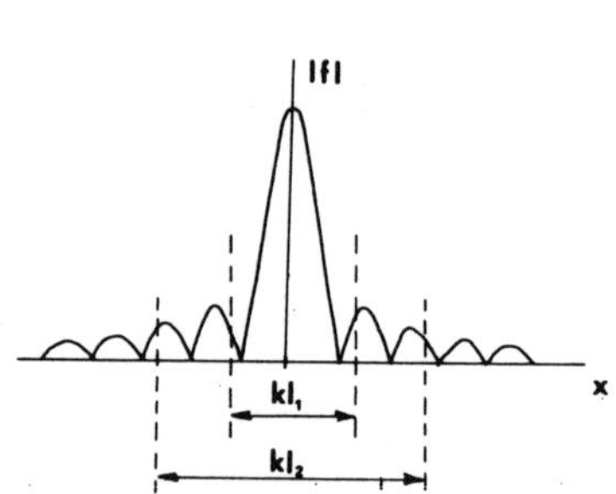

Fig. 9. *Influence of antenna length.*

(c) Influence of phase shift

When the value of p is modified, the maximum position of the $(\sin x)/x$ curve is displaced horizontally. This corresponds to the angle θ_M such that $\cos\theta_M = p$. While $|p| \leqslant 1$, the maximum remains in the visible domain. In the diagram the direction of the maximum, initially $\theta = \pi/2$, slopes onto the axis Oz in a variable direction, according to the sign of p. When $|p| < 1$, the characteristic remains toroidal. For $|p| = 1$, the characteristic corresponds to longitudinal radiation—provided that the characteristic of the elementary source permits it. When, finally, $|p| > 1$, the maximum is in the invisible domain. The value $p = 1$ corresponds to a phase velocity of the travelling waves along the antenna equal to that of light in the ambient medium (Figure 10).

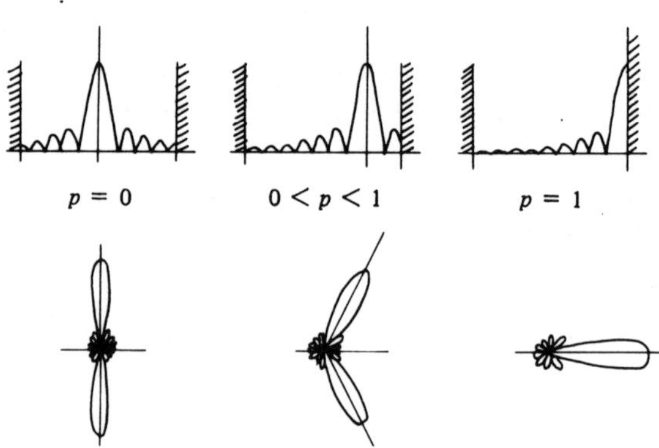

Fig. 10. *Influence of phase shift.*

In the case where $l/\lambda \gg 1$, the position θ_0 of the first minimum is close to the position θ_M of the maximum. So we may write $\theta_0 = \theta_M + \Delta\theta$, with $\Delta\theta \ll \theta_M$.

This gives immediately:

$$\cos \theta_0 = p - \frac{\lambda}{l} = \cos \theta_M - \frac{\lambda}{l} \tag{10}$$

In addition:

$$\cos \theta_0 = \cos (\theta_M + \Delta\theta) \simeq p\left(1 - \frac{\Delta\theta^2}{2}\right) - \Delta\theta\sqrt{1 - p^2} \tag{11}$$

The width of the lobe $\Delta\theta$ is deduced from this as a function of p, using the equation:

$$p\,\Delta\theta^2 + 2\sqrt{1 - p^2}\,\Delta\theta - \frac{2\lambda}{l} = 0 \tag{12}$$

In the two extreme cases of transverse radiation ($p = 0$) and longitudinal radiation ($p = 1$), the lobe widths are:

$$\begin{aligned} \Delta\theta(p = 0) &= \lambda/l \\ \Delta\theta(p = 1) &= \sqrt{2\lambda/l} \end{aligned} \tag{13}$$

which shows that the lobe is narrower in transverse than in longitudinal radiation.

(d) Influence of an amplitude taper

In the case where $l > \lambda$, the alignment characteristic presents, whatever the value of p, a sidelobe level equal to -13.3 dB. For many applications, this level is inadequate, and a distribution characteristic $J(z)$ can be found such that f has no sidelobes. A gaussian-type distribution can evidently be envisaged, the Fourier transform of which would be likewise of gaussian type and therefore would have no sidelobes. But it would then be necessary to have an antenna of unlimited length. Practically speaking, the antenna is truncated, and this results in sidelobes, the amplitude of which are sometimes considerably reduced relative to those of uniform distribution. This reduction in sidelobe level is generally accompanied by a broadening of the principal lobe.

(e) Calculation of directivity

The directivity of the alignment can be calculated using the general relationship:

$$D(\theta_0, \varphi_0) = 4\pi \frac{F'^2(\theta_0, \varphi_0)}{\displaystyle\int_0^{2\pi} d\varphi \int_0^{\pi} F'^2(\theta, \varphi) \sin\theta\, d\theta} \tag{14}$$

Thus, taking (2) into consideration:

$$D(\theta_0, \varphi_0) = 4\pi \frac{F_e'^2(\theta_0, \varphi_0)|f(\cos\theta_0)|^2}{\displaystyle\int_0^{2\pi} d\varphi \int_0^{\pi} F_e'^2(\theta, \varphi)|f(\cos\theta)|^2 \sin\theta\, d\theta} \tag{15}$$

This expression shows that it is generally difficult to express the directivity of the antenna as a function of the directivities of the elementary sources and of that deduced from the alignment function. This becomes possible, however, when it can be assumed that the characteristic F'_e varies little with θ in the area of the alignment function maximum. This situation is encountered particularly for high l/λ ratios. It is possible to establish that:

$$D(\theta_0, \varphi) \simeq \frac{F'^2_e(\theta_0, \varphi_0)}{\frac{1}{2\pi}\int_0^{2\pi} F'^2_e(\theta_0, \varphi)\, d\varphi} \cdot \frac{|f(\cos\theta_0)|^2}{\frac{1}{2}\int_0^{\pi} |f(\cos\theta)|^2 \sin\theta\, d\theta} \tag{16}$$

$$D(\theta_0, \varphi_0) \simeq D_e(\theta_0, \varphi_0)\cdot d_i(\theta_0) \tag{17}$$

where D_e designates the directivity of the elementary source and d_i the directivity of the alignment which would be that of the antenna with isotropic point-sources. If the elementary source has rotational symmetry, then $D_e(\theta_0, \varphi_0) = 1$.

With the introduction of the variable α defined in (5), the directivity d_i in direction α_0 is expressed as:

$$D_i(\alpha_0) = d_i\left(\frac{\alpha_0}{kl}\right) = \frac{4\pi l}{\lambda} \frac{|g(\alpha_0)|^2}{\int_{-kl}^{+kl} |g(\alpha)|^2\, d\alpha} \tag{18}$$

For a distribution of current $J(z) = J_0\, e^{-i\beta z}$, the alignment function is

$$g(\alpha) = \int_{-\infty}^{+\infty} |\mathscr{J}(\xi)|\, e^{i(\alpha\xi - \beta l\xi)}\, d\xi \tag{19}$$

If the directivity is to be calculated in the direction of the radiation maximum, then $\alpha_0 = \beta l$ and:

$$g(\alpha_0) = \int_{-\infty}^{+\infty} |\mathscr{J}(\xi)|\, d\xi \tag{20}$$

By using Parseval's relationship between g and $\mathscr{J}$, it follows that:

$$|g(\alpha_0)|^2 = \int_{-\infty}^{+\infty} |\mathscr{J}(\xi)|^2\, d\xi = \frac{1}{2\pi}\int_{-\infty}^{+\infty} |g(\alpha)|^2\, d\alpha \tag{21}$$

An expression for directivity is deduced from this, involving only the alignment function:

$$D_i(\alpha_0) = \frac{2l}{\lambda} \frac{\int_{-\infty}^{+\infty} |g(\alpha)|^2\, d\alpha}{\int_{-kl}^{+kl} |g(\alpha)|^2\, d\alpha} \tag{22}$$

When l/λ tends towards infinity, the ratio of integrals tends towards unity:

$$D_i(\alpha_0) \simeq \frac{2l}{\lambda}\frac{l}{\lambda} \gg 1 \tag{23}$$

This ratio can be greater than unity for small values of l/λ (Figure 11). It is enough, in fact, for the alignment function to extend mainly into the invisible domain. On the other hand, if the alignment function is entirely located in the visible domain, then the ratio cannot be greater than unity.

It is possible to plot the curve giving the directivity as a function of p for various values of the ratio l/λ in the case of a uniform aperture $J_0(z) = J_0$ (Figure 12). It will be noted that maximal directivity is reached for values of p slightly greater than 1, which is explained by the fact that a slight extension beyond p leads to a large reduction in the width of the principal lobe, or else by the fact that the part of f located in the invisible part is very considerably increased. The optimal value of p has been calculated for a uniform distribution in the case of $l \gg \lambda$:

$$p_{\text{opt}} = 1 + \frac{1}{2l/\lambda} \tag{24}$$

which corresponds to a total shift $\beta l = kl + \pi$. For shorter antennas, the additional shift is from π to $\pi/3$ for $l = \lambda$.

All of the above discussion could be applied to an arbitrary distribution of current. This would show that the value calculated previously gives a maximum and that the directivity is maximal for a distribution of current of uniform amplitude and constant phase characteristic.

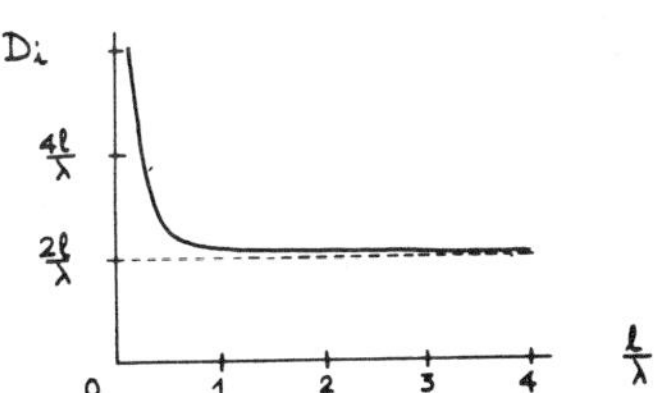

Fig. 11. *Variation of D_i as a function of l/λ for the case of transverse radiation ($p = 0$).*

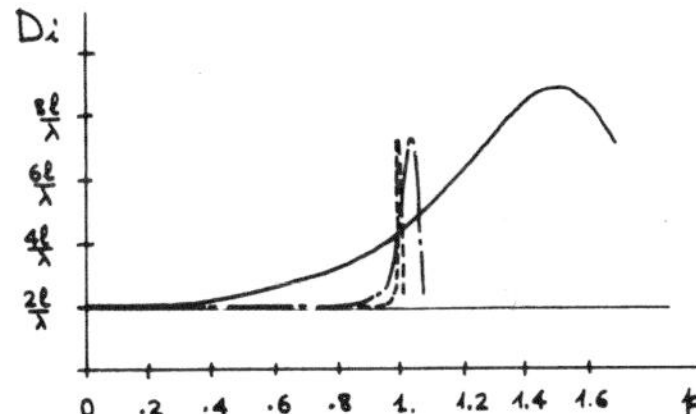

Fig. 12. *Variation in directivity as a function of p, for several values of the ratio l/λ.* (From Walter, *Traveling-wave Antennas*, McGraw-Hill).—$l/\lambda = 1$; —·—· $l/\lambda = 10$; --- $l/\lambda = 100$.

(f) Examples

Travelling wave antennas can be in various forms. They generally have in common the property of having a length of the order of several wavelengths. The distribution of current is obtained by propagation of an electromagnetic wave along a suitable structure: waveguides for fast waves, $0 < p < 1$, surface-wave or wire structures for slow waves, $p \geqslant 1$. The modification of the propagation constant along the structure allows 'electronic scanning' antennas to be made.

The antennas consist of wires or thin slots in metal planes. In the latter case, for the sake of simplicity, the screen will be assumed to be infinite so that the field can be deduced simply from the magnetic currents in the slots.

Example 1: Wire antennas

A large number of antennas are in the form of more or less complex assemblies of taut wires—V-shaped or rhombic antennas, for example. These wires are generally terminated in a suitable load such that it can be assumed, for the purposes of preliminary approximation, that on each of them there is a distribution of the form described by equation (8) (Figure 13). On each of the wires, the wave is propagated with a constant more or less equal to $k(p = 1)$. There cannot be any longitudinal radiation, however, because of the longitudinal aspect of the current. The radiated field presents a maximum in direction θ_m such that the quantity

$$\frac{d}{d\theta}\left\{\sin\theta\frac{\sin\left[\frac{kl}{2}(\cos\theta-1)\right]}{\frac{kl}{2}(\cos\theta-1)}\right\}_{\theta=\theta_m}=0 \tag{25}$$

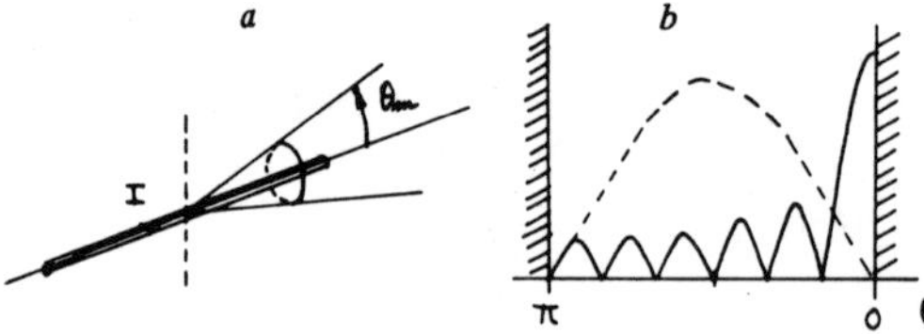

Fig. 13. (*a*) *Wire antenna with travelling wave.* (*b*) *Alignment function* (—) *and elementary dipole characteristic* (– – –).

Putting $y = \sin^2\frac{\theta_m}{2}$, the following transcendental equation is found:

$$\tan y = 2y\left(1-\frac{y}{kl}\right) \tag{26}$$

as long as $y \ll kl$—which is the case for large antennas—$y = 1.165$ rad is obtained. This gives the value of θ_m:

$$\sin\frac{\theta_m}{2} = 0.43\sqrt{\frac{\lambda}{l}} \tag{27}$$

By assembling several wires, a directive antenna can be obtained. The length and orientation of the wires and their height above the ground are important parameters (rhombic antenna).

Example 2: Slot antenna

Consider a slot made in a conductive plane, supplied by a TE_{10} wave from a rectangular guide. If the slot is made on the narrow wall of the guide, it is equivalent to a longitudinal distribution of magnetic currents (Figure 14). In fact, a slot of this type cuts the lines of current.

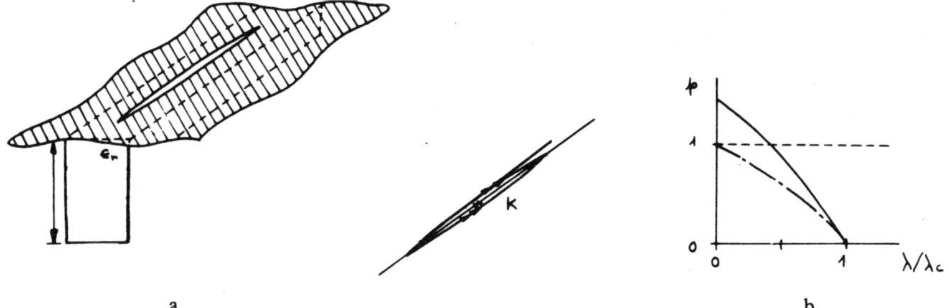

Fig. 14. (*a*) *Slot antenna and equivalent magnetic currents.* (*b*) *Variations of p with* λ ($\lambda c = 2a$). $-\cdot-$ $\varepsilon_r = 1$; — $\varepsilon_r > 1$.

There is an accumulation of charges around the sides of the slot, and a tangential field is created perpendicular to the slot. If the thickness of the slot is small, then for a preliminary approximation it can be accepted that $\beta = \beta_g$—the propagation constant of the closed guide—and that the wave is propagated without attentuation. It is possible, in principle, to vary p between 0 and a maximal value greater than unity (dielectric filling). There are limitations, however; for low values of p, the guide operates around cutoff and the wave is rapidly attenuated; for values greater than or equal to 1, there can be no longitudinal radiation because of the longitudinal character of the currents. From a practical point of view, θ_M can lie between 10° and 80°.

To reduce the sidelobe level, a tapering profile is given to the slot, thus allowing a largely gaussian distribution to be obtained.

If the guide is not terminated at its extremity, the reflected wave gives rise to a diagram similar to that of the direct wave. The comparison of the sidelobe levels gives a measurement of the reflection coefficient.

Note finally that because of the presence of the metallic plane, the radiated field extends only into a half-space.

4: Standing-wave antennas

(a) General points

This section deals with antennas for which the distribution of currents results in standing waves. The distribution is practically equiphasal. In principle, there is no structural difference between travelling-wave antennas and standing-wave antennas. The first can be changed to the second by removing the matching element (terminating loads for V-shaped antennas, matched loads for guides that excite radiating slots) or, better, by replacing them with a short circuit. In fact, standing-wave antennas are distinguished by their length, which rarely exceeds the wavelength.

Behaviour as a function of frequency is quite different in the two types of antenna. For standing-wave antennas, the distribution of the current results from interference, and under these conditions it is clearly strongly influenced by the frequency (appearance or displacement of nodes or antinodes in the current). This sensitivity to frequency has an effect both on the radiated field and the input impedance.

It is generally necessary to work with a particular frequency where the distribution of current is favourable both to the appearance of the radiation characteristic and to the input impedance values. This situation is encountered at so-called resonance frequencies.

The distribution of currents depends largely on the location of the excitation zone. For this reason, a general discussion on this topic is not possible. Only the example of the electric dipole and the structures deriving from it will be considered, including the monopole above a conductive plane and the resonant slot.

(b) The electric dipole

This is a structure made up of two rectilinear wires of the same length $l/2$, collinear and fed by a generator. A more complex theory is combined with experiment to describe the approximate distribution of current in the form:

$$J(z) = J_0 \sin\left\{k\left(\frac{l}{2} - |z|\right)\right\} \tag{28}$$

From this the total current $I(z) = 2\pi a J(z)$ is deduced, where a is the radius of the wires. Note that the distribution of current would correspond to that on an open circuit line with length equal to $l/2$. This analogy is risky, however, and should not be taken any further without great caution.

The electric dipole is therefore a linear alignment of longitudinal dipoles: $\vec{e}_s = \vec{e}_z$. For $\vec{F}_e(\vec{u})$, therefore, the characteristic of this type of dipole is localized at the origin:

$$\vec{F}_e(\vec{u}) = \frac{ik\eta}{4\pi} I(0) l \sin\theta \vec{e}_\theta \tag{29}$$

The field is polarized rectilinearly and is contained in the meridian plane containing the observation point. The calculation of f is easy:

$$f(\theta) = \frac{2}{kl} \frac{\cos\left(\frac{kl}{2}\cos\theta\right) - \cos\left(\frac{kl}{2}\right)}{\sin^2\theta} \tag{30}$$

The norm $F'(\theta)$ of the antenna characteristic is expressed as:

$$F'(\theta) = \frac{I(0)\eta}{2\pi} \frac{\cos\left(\frac{kl}{2}\cos\theta\right) - \cos\left(\frac{kl}{2}\right)}{\sin\theta} \tag{31}$$

The relative characteristic is easier to use:

$$\hat{F}'(\theta) = F'(\theta)/F'(\pi/2)$$

while $F'(\pi/2) \neq 0$, which corresponds to the majority of practical cases ($l < \lambda$). There is no longitudinal radiation whatsoever. It is easy to prove that when the ratio l/λ is varied, the following results are observed:

- When $kl \to 0$, $f(\theta) \to \dfrac{kl}{2}$ and $\vec{F}(\vec{u}) = \dfrac{kl}{2}\,\vec{F}_e(\vec{u})$. This gives, except for a multiplicative factor, the radiation characteristic of a Hertzian doublet (elementary source). This factor reveals the difference between the two current distributions. For the doublet, current I_0 is assumed constant along its length l. For the standing-wave antenna, however, the current distribution is virtually linear. The input current is:

$$I(0) = I_0 \frac{kl}{2}; \quad kl \ll 1 \tag{32}$$

An average current I_m corresponds to it such that:

$$I_m = \frac{1}{l} \int_{-l/2}^{+l/2} I(z)\, dz = I_0 \frac{kl}{4} = \frac{I(0)}{2} \tag{33}$$

The factor $kl/2$ therefore appears as the ratio of the average current in the antenna to the doublet current. The form of the current distribution, therefore, has no influence on the angular dependence of the characteristic. It is involved only at the level of the antenna moment. The importance of charging an antenna capacitively, so as to increase the average current, is clear.

These considerations lead to the definition of an effective length for small antennas. It is, by definition, the length of the Hertzian doublet that, when a certain reference current flows, would create the same field as that on the antenna.

If the average current on the antenna is chosen as reference, it follows that the effective length is equal to the real length of the antenna.

- While $l < \lambda$, the radiation characteristic has only a single lobe in the transverse direction corresponding to $\theta = \pi/2$. The width of the lobe decreases as l increases.
- For $l > \lambda$ sidelobes appear, since the function $F'(\theta)$ can be cancelled out in the directions defined by:

$$\cos \theta_0 = \pm 1 + \frac{4n\pi}{kl} \tag{34}$$

Cancellation takes place in the direction $\theta = \pi/2$ for antenna lengths such that $kl = 4n\pi$ (Figure 15). When moving from one lobe to another, there is a change of phase of π.

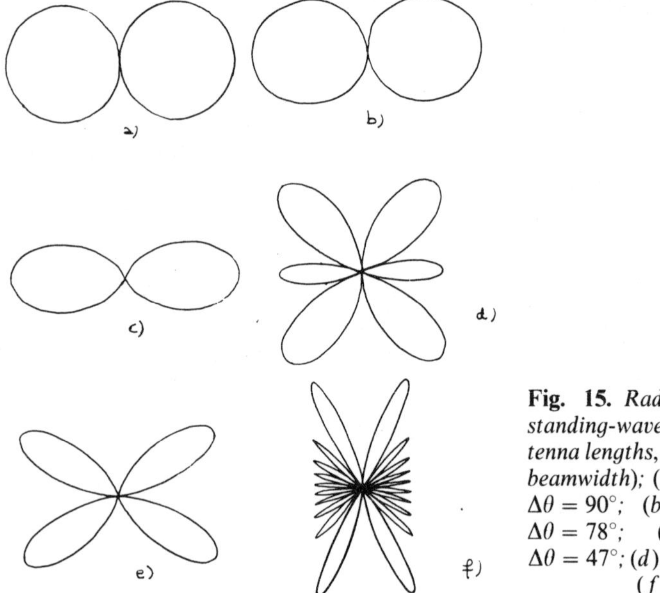

Fig. 15. *Radiation diagrams for standing-wave antennas.* (L = *antenna lengths*, D *directivity*, $\Delta\theta$ *3 dB beamwidth*); (*a*) $L = 0.01\lambda D = 1.5$; $\Delta\theta = 90°$; (*b*) $L = 0.5\lambda D = 1.65$; $\Delta\theta = 78°$; (*c*) $L = \lambda D = 2.76$; $\Delta\theta = 47°$; (*d*) $L = 1.5\lambda$; (*e*) $L = 2\lambda$; (*f*) $L = 10\lambda$.

As already explained, the calculation of the input impedance is difficult. The induced electromotive force method gives satisfactory results for $l \leqslant \lambda/2$. Many other methods have been proposed for the calculation of the input impedance, which will not be summarized here. These methods generally relate to the dipole. The solution to the radiation problem gives easy access to the input impedance.

(c) Structures derived from the dipole

The monopole derives directly from the dipole: it is a wire radiating above a perfectly conductive half-plane. Because of the image principle the monopole can be replaced, above the plane, by a dipole. The distribution of currents is identical. The input impedance is reduced by half. Supply to the monopole is often through a coaxial structure.

Another structure can be deduced from the dipole: the resonant slot. Excitation of the slot takes place directly, for example via a coaxial cable. As a preliminary approximation, it can be stated that a tangential distribution of magnetic currents is established in the slot, such that:

$$\vec{M}(z) = \vec{M}_0 \sin\left\{k\left(\frac{l}{2} - |z|\right)\right\} \tag{35}$$

This relates to the dual structure of the electric dipole. The slot, in an unlimited plane, can be replaced by a distribution *in vacuo*. The input conductance would be deduced from the input resistance by means of the general relationship (44) in the previous chapter.

5: Discontinuous distribution

(a) Calculation of the array factor

The general formula (2) is particularly applicable to discrete dipole distributions:

$$J(z) = \sum_{p=1}^{N} J_p \, \delta(z - z_p) \tag{36}$$

This case is important in practice, since it corresponds to a linear array. The following specific conditions will be adopted:

- The array is regular: $z_p = -\dfrac{1}{2} + (p - 1)d$, where d denotes the distance between two successive elements.
- The array is uniform in amplitude: $J_p = J_1 \, e^{j(p-1)\delta}$, where δ denotes the phase shift between two successive elements.

Relationship (4) remains valid when the currents are normalized in relation to J_1. It follows that:

$$f(\cos\theta) = e^{ikl/2\cos\theta} \sum_{p=1}^{N} J_p \, e^{i(p-1)kd\cos\theta} \tag{37}$$

or:

$$f(\cos\theta) = \frac{\sin\left\{\dfrac{N}{2}(kd\cos\theta - \delta)\right\}}{\sin\left\{\dfrac{1}{2}(kd\cos\theta - \delta)\right\}} \, e^{i\{(N-1)[(kd/2)\cos\theta - \delta] - (kl/2)\cos\theta\}} \tag{38}$$

The modulus of f is of particular interest:

$$|f(\cos\theta)| = \left|\frac{\sin NX}{\sin X}\right| \quad \text{with} \quad X = \frac{1}{2}(kd\cos\theta - \delta) \tag{39}$$

The same result would be obtained by using the translation theorem. By taking $\vec{E}_1(r, \vec{u})$ as the field created by the first antenna at distance r, in direction u, the expression for the total field is obtained without difficulty, in the form:

$$\vec{E}(r, \vec{u}) = \vec{E}_1(r, \vec{u}) \frac{\sin NX}{\sin X} e^{i\psi}$$

where ψ is the argument that appears in formula (38).

(b) Discussion

The difference between the function f relative to a discontinuous distribution—often called the array factor—and that relative to a continuous distribution is in the denominator, which is here a periodic function of X, and therefore of $\cos\theta$ in the visible domain (Figure 16). The same function repeats periodically, with the period $X = \pi$ for $|f|$.

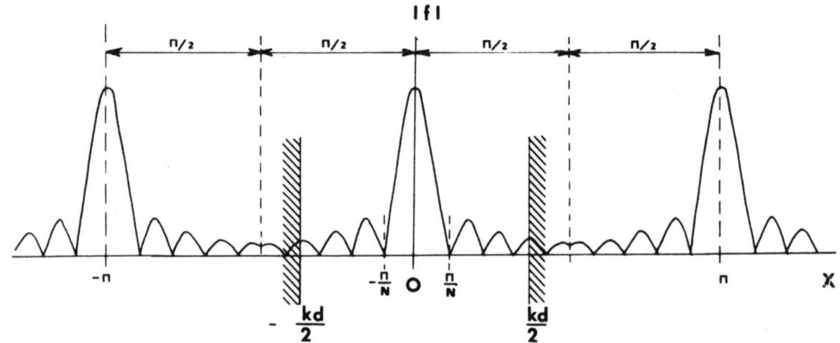

Fig. 16. *Array factor.*

For $\delta = 0$, the visible domain is limited to the interval $-\dfrac{kd}{2}, +\dfrac{kd}{2}$. Depending on the value of the ratio λ/d, there may or may not be several principal lobes in the visible zone. Within a single period, there is a pseudo-periodicity of $X = \pi/N$. The width of the various lobes is fixed by $ND \simeq l$, whereas the repetition period of the function is independent of N and depends only on d. When $d < \lambda/2$, there is only one principal lobe in the visible domain and the level of the sidelobes decreases monotonically. If $d > \lambda$, there are at least two principal lobes. The first situation arises, for example, with radar antennas, for which there must be no ambiguity in the direction of the target. The second is encountered in interferometric baselines used in radio astronomy: the array is then made up of two antennas, the distance between which can be modified, thus allowing the ambiguity resulting from the existence of several lobes to be resolved.

The phase shift δ fixes the position of the function f in the visible domain. In this way, it is possible to control the shift of various elements to make the principal lobe scale. This is the principle behind array antennas with electronic scanning. For $\delta = 0$, the array radiates in a broadside direction. For $\delta = \pi$, the radiation is longitudinal.

When N tends towards infinity such that Nd = constant (array of constant total length), and $N\delta$ = constant (constant total shift), the array factor can be expressed in the form:

$$|f(Y)| = N\left|\frac{\sin Y}{Y}\right| \quad \text{with} \quad Y = NX \tag{40}$$

When N increases, the main lobe -3 dB beamwidth decreases and the sidelobe level tends towards the -13.3 dB of the continuous excitation. At the limit $N \to \infty$, the characteristic function of the continuous excitation is found.

Note that for a real array, the radiation characteristic is the product of the array factor and the radiation characteristic of one of the array elements.

There is generally a resulting rapid reduction in the maxima situated in the invisible domain.

(c) Calculation of directivity

For any number N of elements, the directivity associated with the array factor can be calculated using the general relationship (14). By using the formula:

$$\left|\frac{\sin NX}{\sin X}\right|^2 = N + 2\sum_{p=1}^{N-1}(N-p)\cos 2pX \tag{41}$$

the expression of directivity in direction θ_0 is obtained:

$$d(\theta_0) = \frac{1}{\dfrac{1}{N} + \dfrac{2}{N^2}\displaystyle\sum_{p=1}^{N-1}\frac{N-p}{pkd}\sin pkd\cos p\delta} \tag{42}$$

Note that the denominator is reduced to the term $1/N$, whatever the shift δ, while kd is an integer multiple of π, with d an odd multiple of the half wavelength. Under these conditions, there is at least one principal lobe in the visible domain. In the corresponding direction, the array factor is l and the directivity is N. This result should be compared with the directivity of a continuous alignment for which $2l/\lambda$ was found. The two results coincide when N is large $l \simeq Nd$ and $d = \lambda/2$.

(d) Examples

Consider the type of array that allows longitudinal radiation to be obtained (unlike the antennas described above). This effect can be achieved by choosing elementary sources the moments of which are perpendicular to the array axis (electric or magnetic dipoles). The magnetic dipoles could be produced with transverse slots made in the longer side of a waveguide. So far, the case of arrays with uniform amplitude has been considered. As for continuous excitations, a trade off of the distribution of amplitude in the elementary sources would allow the compromise between the width of the principal lobe and the sidelobe level to be improved. This effect can be obtained either with a regular array, the amplitude of which would not be constant, or with an irregular array in which the amplitude of each element would be constant.

Finally, it should be noted that all discussion of the array factor is also valid for arrays made up, not of elementary sources, but of identical antennas. The only approximations made concerning the antennas relate to mutual coupling: it is generally assumed that the radiation characteristic of an antenna, in the presence of others, is equal to its characteristic when it is alone, and whatever its position in the array (edge effects).

C: Planar structures

1: General formulae

These concern plane apertures S, with dimensions that are large compared to the wavelength. This configuration occurs principally at microwave frequencies, in many antennas, the structures of which are not otherwise necessarily planar (lenses, reflectors, etc). As with linear structures, there can be a continuous or discontinuous distribution.

Only continuous distributions will be considered here. Rather than reasoning, as before, on the basis of the moments of electric or magnetic dipoles, it is more convenient to bring in the values of the tangential components of the electromagnetic field directly. Consider an electric field polarized uniformly along axis Ox (Figure 17):

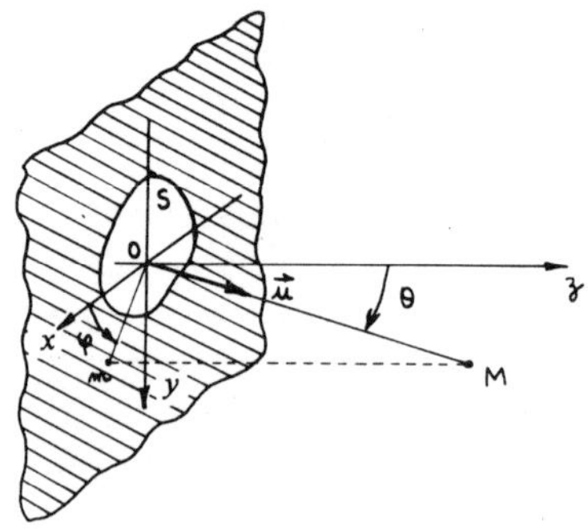

Fig. 17. *Radiating aperture.*

$$\vec{E}(r') = E(r')\vec{e}_x; \quad r' \in S \tag{43}$$

If the equivalent sources lie on a plane, the field will be a function only of the tangential electric field, with the relationship:

$$\vec{E}(\vec{r}, \vec{u}) = \frac{e^{-ikr}}{r} \vec{F}'(\vec{u}) \tag{44}$$

where:

$$\vec{F}'(\vec{u}) = \frac{ik}{2\pi} \iint_S E(r')\, e^{ik\vec{r}'\cdot\vec{u}}\, dr'\vec{e}' \tag{45}$$

$$\vec{e}' = \cos\theta \cos\varphi \vec{e}_\theta - \sin\varphi \vec{e}^\varphi \tag{46}$$

A second relationship can be derived by taking account of the contribution of the tangential magnetic field. A locally plane wave structure in the aperture is often admitted, such that:

$$\vec{H}(\vec{r}') = \frac{1}{\eta} \vec{e}_z \wedge \vec{E}(r') \tag{47}$$

This give a radiation characteristic:

$$\vec{F}''(u) = \frac{ik}{4\pi}(1 + \cos\theta) \iint_S E(r')\, e^{ik\vec{r}'\cdot\vec{u}}\, dr'\vec{e}'' \tag{48}$$

$$\vec{e}'' = \cos\varphi \vec{e}_\theta - \sin\varphi \vec{e}_\varphi \tag{49}$$

Other expressions can also be obtained. Note that the two previous relationships are practically equivalent around the direction $\theta = 0$:

$$\vec{F}(\vec{u}) = \vec{F}'(\vec{u}) \simeq \vec{F}''(\vec{u}) \\ = f(\alpha, \beta)\vec{F}_m(\vec{u}) \tag{50}$$

with:

$$\vec{F}_m(\vec{u}) = \frac{ik}{4\pi} 2E(0, 0)\vec{e}' \tag{51}$$

$$f(\alpha, \beta) = \iint_S \frac{E(x', y')}{E(0, 0)} e^{ik[\alpha' x + \beta y']} dx'\, dy' \tag{52}$$

and α, β, γ designating the direction cosines in the direction $\vec{u}$ in system $Oxyz$. If the direction is given in spherical coordinates, then:

$$\begin{aligned} \alpha &= \sin\theta \cos\varphi \\ \beta &= \sin\theta \sin\varphi \\ \gamma &= \cos\theta \end{aligned} \tag{53}$$

The characteristic $\vec{F}$ is therefore the product of the characteristic $\vec{F}_m$ of a magnetic dipole of moment equal to $2E(O, O)$ parallel to Oy and located at the origin by the alignment function $f(\alpha, \beta)$. This function therefore appears as the Fourier Transform of the illumination or distribution law in the aperture, thus generalizing the property encountered for linear structures.

For antennas where $S \gg \lambda^2$, the diagram is concentrated in a restricted solid angle in which the radiation characteristic $\vec{F}_m$ can be considered practically constant.

2: Examples

(a) Separable apertures

For this type of aperture, variables x and y can be separated:

$$E(x, y) = E_1(x)E_2(y) \tag{54}$$

There is immediate corresponding separation in:

$$f(\alpha, \beta) = f_1(\alpha)f_2(\beta) \tag{55}$$

The functions f_1 and f_2 can be interpreted as alignment functions of sources located along Ox and Oy. In the plane $\varphi = 0$ (E-plane), $\beta = 0$ and therefore:

$$f(\alpha, 0) = f_1(\alpha)f_2(0) \tag{56}$$

There is a similar property in the plane $\varphi = \pi/2$ (H-plane) for which $\alpha = 0$. The application to the rectangular aperture, of dimensions a, b along the respective axes Ox and Oy, gives the case of uniform illumination:

$$f(\alpha, \beta) = ab \frac{\sin \frac{k\alpha a}{2}}{\frac{k\alpha a}{2}} \cdot \frac{\sin \frac{k\beta b}{2}}{\frac{k\beta b}{2}} \tag{57}$$

In E- and H-planes the sections of the radiation diagram admit, for the width of the principal lobe, the half-angles θ_E and θ_M such that:

$$\sin \theta_E = \frac{\lambda}{a} \qquad \sin \theta_M = \frac{\lambda}{b} \tag{58}$$

If a and b are large compared to λ, the sine becomes approximately equal to its argument: $\sin \theta_E \approx \theta_M$ and $\sin \theta_M \approx \theta_M$.

(b) Circular aperture

Circular apertures are often encountered in practice. The calculation is easy to carry out in polar coordinates. The function f is independent of φ; for uniform illumination, the following is obtained:

$$\begin{aligned} f(\alpha, \beta) &= \int_0^{2\pi} d\varphi' \int_0^{a/2} e^{ikr' \sin\theta \cos(\varphi - \varphi')} r' \, dr' \\ &= \frac{\pi}{2} a^2 \frac{J_1\left(\frac{ka}{2} \sin \theta\right)}{\frac{ka}{2} \sin \theta} \end{aligned} \tag{59}$$

where a = diameter, J_1 = Bessel function.

The first cancellation takes place for a value θ_0, such that:

$$\sin \theta_0 = 1.22 \frac{\lambda}{a} \tag{60}$$

3: Non-uniform apertures

The aperture with uniform amplitude and constant phase produces a field maximum radiated in the direction $\theta = 0$. The level of the first sidelobe is of the order of 13 dB below the level of the principal lobe. As for linear sources, tapering of the illumination characteristic on the edges of the aperture (apodization) allows a reduction in sidelobe level at the cost of an increase in the 3 dB beamwidth and a reduction in gain. This technique is widely used in practice: the choice of the primary source to illuminate a parabolic reflector generally takes this into account. The diagram of this source shows, in the directions corresponding to the edges of the reflector, an attenuation of the order of some 10 dB. Specific distributions allow sidelobe levels of about 20 dB below the level of the principal lobe.

The effect of a linear phase characteristic is important:

$$E(x, y) = E_0\, e^{-(k\alpha' x + k\beta' y)} \tag{61}$$

The Fourier shift theorem allows the function f of this type of distribution to be expressed immediately in terms of the function describing the uniform and equiphasal aperture:

$$f(\alpha, \beta) = f_0[(\alpha - \alpha'), (\beta - \beta')] \tag{62}$$

There is a rotation of the direction of the principal lobe: this now lies in the direction defined by the direction cosines α', β' and $\sqrt{1 - \alpha'^2 - \beta'^2}$. For large apertures, these direction cosines are directly proportional to the angle of rotation.

Consider, finally, the case of a distribution with uniform amplitude, displaying a quadratic phase characteristic. This situation could be encountered with a reflector illuminated by a primary source, shifted axially from the theoretical focus position. The result is generally a degradation in sidelobe level and an increase in 3 dB beamwidth. A useful application of this will be seen in the chapter on measurement in the simulation, at finite distance, of the radiation characteristic.

4: Calculation of directivity

The general formula (24) from Chapter 4 is applied to radiating apertures. In direction $\vec{u}$, the density of power $\sigma(\vec{u})$ is:

$$\sigma(\vec{u}) = \frac{1}{2\eta} E'^2 r^2 = \frac{1}{2\eta}\frac{k^2}{4\eta^2}\left|\iint_S E(x', y')\, e^{ik(\alpha x' + \beta y')}\, dx'\, dy'\right|^2 \tag{63}$$

The calculation of the active radiated power W'_r can be achieved using the flow of Poynting's vector through the aperture (Chapter 4, equation (17)):

$$W'_r \operatorname{Re}\left\{\iint_S \vec{P}\cdot\vec{n}\, dS\right\} = \frac{1}{2\eta}\iint_S E'^2(x', y')\, dx'\, dy' \tag{64}$$

The directivity $D(\vec{u})$ can be deduced straight away:

$$D(\vec{u}) = \frac{4\pi}{\lambda^2}\frac{\left|\iint_S E(x', y')\, e^{ik(\alpha x' + \beta \gamma')}\, dx'\, dy'\right|^2}{\iint_S E'^2(x', y')\, dx'\, dy'} \tag{65}$$

The directivity is maximal in the direction $\alpha = \beta = 0$, that is $\theta = 0$, with an equiphasal aperture. So, according to Schwartz's inequality:

$$\left|\iint_S E(x', y')\, e^{ik(\alpha x' + \beta y')}\, dx'\, dy'\right|^2 \leqslant \iint_S E'^2(x', y')\, dx'\, dy' \iint_S dx'\, dy' \tag{66}$$

Equality occurs under the conditions cited previously:

$$D(\alpha, \beta) \leqslant D(\mathrm{O}, \mathrm{O}) = \frac{4\pi S}{\lambda^2} \tag{67}$$

The maximal value of directivity for a given aperture with surface S therefore follows immediately. The general formula (65) allows the calculation to be made for an arbitrary distribution. Directivity is expressed in the form:

$$D(\alpha, \beta) = g\frac{4\pi S}{\lambda^2} \tag{68}$$

where $g \leqslant 1$ denotes the gain or directivity factor.

Chapter 6

Introduction to antenna synthesis

A: General points

So far in this work, antennas have been considered from an analytical point of view, in order to calculate the field radiated by a known distribution of currents. The intention in this chapter is to tackle the problem of synthesis, which is more important in practice, and consists of determining the currents that are likely to radiate, at great distances, a far field with certain properties defined in advance.

The problems of synthesis differ greatly, depending on the way in which the properties to be obeyed in the far field are defined, and the way in which the currents are parameterized. The field can be defined by its radiation characteristic in all or part of the visible domain—that is, more specifically, for the values of this characteristic in a certain number of directions. It can also be defined by overall values, such as the 3 dB beamwidth and/or the sidelobe level. The use of such values generally leads to a compromise, since it is not possible to impose over-strict constraints on all the parameters. Synthesis therefore involves optimization methods, the cost function of which is defined simply on the basis of the performance of the antenna, or of the system in which the antenna functions. Very often, the antenna will operate with directive radiation, or else a narrow principal lobe and a low sidelobe level.

The synthesis method also depends on the parameterization of the currents, the support of the antenna and mode of representation which must be specified.

In general, current forms are chosen to have easily calculated radiation characteristics, which can be reproduced in practice on the chosen antenna. This is important in synthesis: there is no point in determining, by calculation, currents that cannot subsequently be realized. This is why, for continuous structures, currents in the form of standing or travelling waves, damped or not, are found. From the point of view of practical realization, discontinuous structures seem to offer a high level of flexibility. In fact, it is not possible to fix arbitrarily the excitation of each element, because of mutual coupling phenomena.

Taking into account the variety of problems that can arise, it is clear that there cannot be one method of synthesis applicable to all situations, but rather a large number of specific methods.

Only those methods applicable to continuous or discontinuous linear structures will be considered here. These methods will illustrate the difficulties that can be encountered in problems of synthesis, and more particularly when an arbitrarily large directivity is to be obtained with an antenna of fixed dimension.

There is no point in summarizing results based directly on the existence of a Fourier Transform between the distribution of currents and the radiation characteristic. If currents are to be determined, the characteristics in the visible and invisible domains must be known. For a real antenna, however, with finite dimensions, the characteristic is an entire function (Fourier Transform of a function with bounded support) and, as such, is theoretically perfectly defined by its values in a limited part of the visible or invisible spectrum. On a more concrete level, it is possible to envisage characteristics that are almost exact in the visible domain and very different in the invisible domain. This will be further discussed below.

From another point of view the idea of a finite antenna is rather vague, in that a structure on which a wave with an exponentially decreasing current is propagated, can be considered as finite, for example. The characteristic of such an antenna is not, however, an entire function and the study of its poles plays an important role, as in the synthesis of circuits.

B: Examples

1: Woodward's method

In this method, the field is characterized by the values of its characteristic, assumed to be real in a certain number of directions. The current, assumed to be continuous, is broken down into travelling waves (space harmonics). In this and the following paragraphs, it will be expressed independently of its electric or magnetic aspect by $A(z)$:

$$\left.\begin{aligned} A_N(z) &= \sum_{n=-N}^{+N} a_n\, e^{-i2\pi n(z/l)} \quad |z| \leqslant \frac{l}{2} \\ A_N(z) &= 0 \qquad\qquad\qquad\quad |z| > \frac{l}{2} \end{aligned}\right\} \tag{1}$$

The problem consists of determining N so that the coefficients a_n, $n = 1, 2, \ldots, N$, for the field radiated by A_n will be as close as possible to the given characteristic $f(\cos\theta)$. The characteristic f_N associated with A_N can immediately be written:

$$\begin{aligned} f_N(\cos\theta) &= \sum_{n=-N}^{+N} a_n \int_{-(l/2)}^{+(l/2)} e^{i[k\cos\theta - 2\pi(n/l)]z}\, dz \\ &= \sum_{n=-N}^{+N} a_n \frac{\sin\left\{\dfrac{kl}{2}\cos\theta - n\pi\right\}}{\dfrac{kl}{2}\cos\theta - n\pi} \end{aligned} \tag{2}$$

f_N appears in the form of a sum of elementary shifted $(\sin x)|x$ type diagrams.

The coefficients a_N can be chosen, for a given N, so as to minimize the error $f - f_N$ in the interval $(0, \pi)$. The average quadratic error (least mean squares method) can therefore be minimized, as can the maximal error (Tchebycheff's method). The simplest solution consists of noting that for all the values of $\cos\theta_n$ such that:

$$\cos\theta_n = \frac{2n\pi}{kl} = \frac{n\lambda}{l} \tag{3}$$

all the terms in the sequence are zero, with the exception of that of rank n. The coefficients a_n can thus be determined from a known f in N directions:

$$a_n = f\left(\frac{2n\pi}{kl}\right) = f_N\left(\frac{2n\pi}{kl}\right) \tag{4}$$

It is possible to envisage increasing the number of sampling directions arbitrarily. When the directions of sampling are augmented, however, or when an antenna is to be obtained with directivity notably greater than $2l/\lambda$, the function f_N admits large values in the invisible domain. This point will be discussed further in section 3.

2: Using the Laplace transform

By introduction of the complex variable $p = p' + ip''$, such that $p'' = k\cos\theta$, the radiation characteristic can be extended to the complex plane. If the antenna is located on the interval (0, 1), it follows that:

$$f(p) = \int_0^\infty A(z)\, e^{pz}\, dz \tag{5}$$

This is easily recognized as a Laplace transform, which by inversion allows A to be determined as a function of f. In the complex plane, the visible domain corresponds to the segment of the imaginary axis $|p''| \leqslant k$. The function f will be chosen so as to respect, in this domain, a certain dimension (cf the analogies with circuit theory)—of the Butterworth or Tchebycheff types, for example. Generally, $f(p)$ can be found in the form of a simple pole development:

$$f_N(p) = \sum_{n=1}^{N} a_n \frac{1}{p - p_n}; \quad p_n' > 0 \tag{6}$$

The current will be in the form of a superposition of damped travelling waves, of infinite extent:

$$A_N(z) = \sum_{n=1}^{N} a_n\, e^{-p_n z} \tag{7}$$

The characteristic f'_N of a truncated antenna, with distribution:

$$A'(z) = \begin{cases} A_N(z) & z \leqslant l \\ 0 & z > l \end{cases} \tag{8}$$

is deduced simply from the characteristic f_N:

$$f'_N(\cos\theta) = f_N(\cos\theta) * \frac{\sin\left\{\dfrac{kl}{2}\cos\theta\right\}}{\dfrac{kl}{2}\cos\theta} \tag{9}$$

3: Dolph–Tchebycheff's method

Unlike the two previous methods, this applies to arrays. For an array with N elements, the characteristic f_N is expressed as:

$$f_N(\cos\theta) = e^{-(ikl/2)\cos\theta} \sum_{n=1}^{N} a_n\, e^{i(n-1)\phi} = g_N(\varphi) \tag{10}$$

with

$$\varphi = kd\cos\theta - \delta$$

For a symmetrical array with an even number of elements:

$$g_N(\varphi) = 2l^{-i(kl/2)\cos\theta}\, e^{il(N-1)(\varphi/2)} \sum_{p=1}^{N/2} a_p \cos(N-p)\frac{\varphi}{2} \tag{11}$$

The sum of the cosines can be identified using a Tchebycheff polynomial of degree $N - 1$:

$$\sum_{p=1}^{N/2} a_p \cos(N-p)\frac{\varphi}{2} = \sum_{p=1}^{N/2} a_p T_{N-p}\left(\cos\frac{\varphi}{2}\right) \tag{12}$$

In certain applications it may be useful to have a constant sidelobe level. The characteristic can be approximated by the polynomial

$$g_N(\varphi) = \frac{T_{N-1}\left(\mu\cos\dfrac{\varphi}{2}\right)}{T_{N-1}(\mu)} \tag{13}$$

where μ is a parameter to be determined. Taking into account the properties of the Tchebycheff polynomials, it is easy to verify that the sidelobe level is constant and equal to:

$$A = \frac{1}{T_{N-1}(\mu)} \tag{14}$$

The first null θ_0 allows the width of the principal lobe to be characterized:

$$\cos\frac{\varphi_0}{2} = \frac{1}{\mu}\cos\frac{\pi}{2(N-1)} \tag{15}$$

With a given sidelobe level A, μ and θ_0 can be determined. Conversely, if θ_0 is imposed, A can be deduced from it. A remarkable property of Tchebycheff's polynomials guarantees that, of all the polynomials of degree $N - 1$, the beamwidth θ_0 obtained is the smallest for a given sidelobe level or, on the other hand, that A is the smallest for a given width θ_0. By identification of (12) and (13), the coefficients a_p can be determined.

Elementary properties of Tchebycheff's polynomials
The definition is summarized here. Tchebycheff's polynomial of order n is defined by:

$$T_n(x) = \cos(n \operatorname{arc} \cos x)$$

For $|x| \leqslant 1$, $x = \cos u$ is set; $T_n(x) = T_n(\cos u) = \cos nu$. This gives $|T_n(x)| \leqslant 1$. For $|x| > 1, x = \pm\cosh u$ is set; $T_n(x) = T_n(\cosh u) = \cosh nu$ grows very rapidly as a function of x.

The polynomial T_n is cancelled out n times in the interval $|x| \leqslant 1$. The cancellations occur for the values x_k such that $x_k = \cos(2k + 1)\dfrac{\pi}{2n}$.

C: Introduction to superdirectivity

As was seen in the previous sections, the size of an antenna has a great deal of influence on its directivity. This is particularly apparent for large antennas ($l/\lambda \gg 1$), the directivity of which is directly proportional to the ratio l/λ. It is less evident for small antennas ($l/\lambda > 1$), the directivity of which tends towards zero because of the rule of proportionality summarized above. A half-wave dipole antenna has almost the same directivity as a Hertzian doublet.

A difference will be noted between the two types of antenna. For large antennas, the radiation characteristic is located in a narrow zone in the visible domain. For small antennas, however, the radiation characteristic extends a considerable way into the invisible domain. A superdirectivity factor R is therefore defined using the relationship:

$$R = \frac{\displaystyle\int_{-\infty}^{+\infty} |g(\alpha)|^2 \, d\alpha}{\displaystyle\int_{-kl}^{+kl} |g(\alpha)|^2 \, d\alpha} \tag{16}$$

Because of equations (22) and (23) from Chapter 4, the maximal directivity of an array of length l is expressed:

$$D_i(\alpha) = R\,\frac{2l}{\lambda} \tag{17}$$

The antenna is said to be superdirective when the factor R is greater than unity, although this is a definition based purely on convention. This situation arises when the radiation characteristic is contained entirely in the visible domain $|\alpha| \leqslant kl/2$. In this respect, the small antennas mentioned above are superdirective.

To return to the problem of synthesis, the first question that arises is whether it is possible to obtain a radiation characteristic with a beamwidth $\Delta\theta$ in the visible domain, using a distribution of sources with an arbitrarily small extent. If this characteristic does not admit any extension into the invisible domain, then the antenna length will have to be of the order of $l \simeq \lambda/\Delta\theta$. If, on the other hand, this characteristic is suitably prolonged in the invisible domain then a factor R may be obtained that is far larger than unity.

The second question is how to arrive at this. It is necessary to create conditions of strong interference between the various elements in the excitation. This gives rise to current distributions that oscillate strongly and have a large amplitude. If no constraint is imposed, other than the beamwidth in the visible domain, or the directivity, it can be shown that it is theoretically possible to obtain an antenna length l as short as is required.

This theoretical possibility cannot, however, easily be applied. Firstly, it is out of the question to obtain the desired distribution in the form of a continuous structure, because of its complexity. Even if a discontinuous structure is envisaged, it will remain very difficult to impose the desired distribution because of mutual coupling phenomena. Secondly, even if this were achieved, there are still grave difficulties. The first consists of a drop in efficiency loss due to the magnitude of the currents that have to flow, bringing about a rapid increase in ohmic losses. The second concerns the selectivity of the antenna, both in terms of the radiated field and of the input impedance. The strong interference conditions between the various elements in an array are very sensitive to variations in the frequency, and the characteristic changes rapidly. In addition, the extension of the characteristic in the invisible domain corresponds to a reactive power accumulated around the antenna. The result is that the input reactance is larger when R is large compared to 1. This may have to be compensated for by a matching element, so as to obtain the resonance. A circuit formed in this way is itself very selective and limits the band over which the antenna can be used.

Note that the existence of a high superdirectivity factor does not appear only when the directivity of an antenna is being increased beyond $2l/\lambda$ but also when, in Woodward's method, the number of samples is increased to improve the 'definition' of the characteristic.

Bibliography to Chapter 6

CHU L. J. Physical limitations of omni-directional antennas, *J. Appl. Phys.*, vol. **19**, December 1948, pp. 1163–1175.

COLLIN R. E., ZUCKER F. *Antenna Theory*, McGraw-Hill 1969, Chap. 7, by SCHELL A. C. and ISHIMARU A. 'Antenna pattern synthesis'.

DOLPH C. L. A current distribution for broadside arrays which optimize the relationship between width and sidelobe level, *Proc. IRE*, vol. **34**, June 1946, pp. 335–348.

HANSEN W. W., WOODYARD J. R. A new principle in antenna design, *Proc. IRE*, vol **26** (3), March 1938, pp. 333–345.

HARRINGTON R. F. *Time-harmonic Electromagnetic Fields*, pp. 307–311, McGraw-Hill, New York 1961.

TAYLOR J. J. A note on the maximum directivity of an antenna, *Proc. IRE*, 1958, p. 620.

WOODWARD P. M., LAWSON J. D. The theoretical precision with which an arbitrary radiation pattern may be obtained from a source of finite size, *J. Inst. Elect. Eng.*, 8t 3, 1948, pp. 363–369.

YARU N. A note on super-gain antenna arrays, *Proc. IRE*, vol. **39**, September 1951, pp. 1081–1085.

Chapter 7

Receiving antennas

A: Introduction to reception theory

The object of this chapter is to characterize a receiving antenna—that is, an antenna subjected to the field radiated by another antenna. This consists of calculating the power supplied by the antenna to the reception system. In relation to this system, the antenna can be represented by an equivalent generator, the electromotive force and internal impedance of which will be related to the corresponding characteristics (gain, input impedance) of the same antenna operating in transmission mode and to those of the illuminating field. This is assumed to be purely monochromatic, with a well-defined state of polarization. This will provide a preliminary approximation for real problems of radio links where the propagation media introduce fluctuations in level (partial coherence) and polarization (partial polarization). The elements of the equivalent generator will be determined here for a plane incident wave, first of all. After this, consideration will be given to the coupling phenomena that arise between the reception and transmission antennas, creating the field of illumination. In this way, exact and general formulae will be determined, which are greatly simplified by the usual hypothesis of weak coupling. In this hypothesis, the reaction of the receiving antenna to the transmitting antenna is ignored, and the latter can be entirely characterized by the field it creates in the absence of the receiving antenna. This hypothesis is very suitable for use in cases where the two antennas are very distant from each other, a situation that often arises in telecommunciations.

Once the elements of the generator equivalent to the antenna have been calculated, there is no difficulty in discussing the influence of different parameters of the problem on the power supplied by the antenna to the reception system. As for any element in a reception chain, it is important to know how the antenna will behave in the presence of noise. Only thermal noise will be discussed here. This results from the incoherent and unpolarized electromagnetic radiation emitted by bodies surrounding the antenna.

B: Antenna response to an incident plane wave

1: Stating the problem

Let Ω (Figure 18) be a domain with boundary $\Sigma = S \cup S_\infty$, where S is a surface surrounding the antenna in question and S_∞ is a sphere of large radius, centred on the origin; $\vec{n}$ is the unit vector normal to Σ and directed out of Ω. The domain Ω is made up of a passive, linear and possibly inhomogeneous and anisotropic medium. Even in the latter case, it is assumed nonetheless that the tensors of permittivity and permeability are symmetrical.

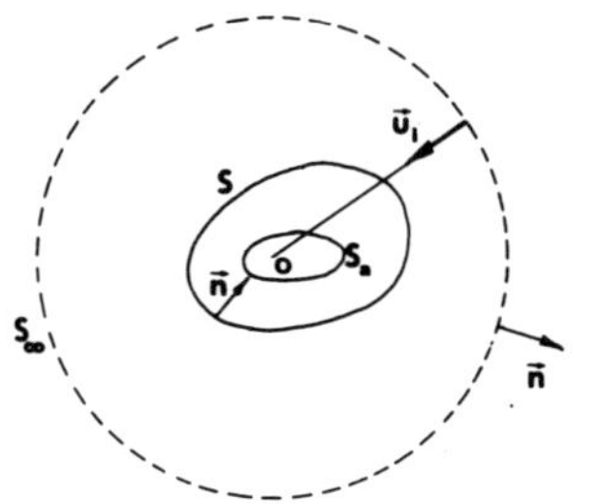

Fig. 18.

Two states are considered, a and b, defined as follows:

State a: The antenna functions in transmission mode. It creates, at all points in the space, a field $\vec{E}^a$, $\vec{H}^a$. In particular, at distance r, in direction $\vec{u}$:

$$\vec{E}^a = \frac{e^{-ikr}}{r} \vec{F}^a(\vec{u}) \tag{1}$$

$$\vec{H}^a = \frac{1}{\eta} \vec{u} \wedge \vec{E}^a \tag{2}$$

where $\vec{F}^a$ is the vector radiation characteristic of the transmitting antenna:

$$\vec{F}^a = \frac{ik}{4\pi} \int_{S_a} \{(\eta \vec{J}^a \wedge \vec{u}) \wedge \vec{u} - \vec{M}^a \wedge \vec{u}\} \, e^{ik\vec{r}' \cdot \vec{u}} \, dr' \tag{3}$$

where $\vec{J}^a$ and $\vec{M}^a$ are the equivalent sources on the antenna surface S_a.

State b: The antenna is illuminated by a plane wave propagating in direction $\vec{u}_i$, the field of which, with arbitrary polarization, is:

$$\vec{E}_i^b = e^{ik\vec{r} \cdot \vec{u}_i} \vec{E}_0 \tag{4}$$

$$\vec{H}_i^b = \frac{1}{\eta} \vec{u}_i \wedge \vec{E}_i^b \tag{5}$$

The presence of the antenna disturbs this distribution: the plane wave induces in the antenna currents that in their turn radiate, a field $\vec{E}_r^b$, $\vec{H}_r^b$. The total field in state b is:

$$\vec{E}^b = \vec{E}_i^b + \vec{E}_r^b \tag{6}$$

$$\vec{H}^b = \vec{H}_i^b + \vec{H}_r^b \tag{7}$$

In the far field the radiated field satisfies the radiation condition, and can be expressed as follows:

$$\vec{E}_r^b = \frac{e^{-ikr}}{r} \vec{F}^b(\vec{u}) \tag{8}$$

$$\vec{H}_r^b = \frac{1}{\eta} \vec{u} \wedge \vec{E}_r^b \tag{9}$$

where $\vec{F}^b$ is the vectorial radiation characteristic associated with the currents induced in the antenna during reception. These currents are usually different from those of transmission mode, so that $\vec{F}^b$ is generally different from $\vec{F}^a$. Consider the integral $I(S)$ defined by:

$$I(S) = \int_S (\vec{E}^a \wedge \vec{H}^b - \vec{E}^b \wedge \vec{H}^a) \cdot \vec{n}\, dS \tag{10}$$

As will be seen, this integral does not depend on the choice of S as long as this surface surrounds the antenna. If S is chosen to coincide with the surface of the antenna S_a, it can be calculated by two different methods, one very general, taking into consideration the properties of the radiated fields, the other, more specific, taking account of their structure on the surface of the antenna.

2: Calculation of $I(S)$

(a) Breakdown of the field

By breaking down the total field in state b, it follows that:

$$\begin{aligned} I(S) &= \int_S [\vec{E}^a \wedge \vec{H}_i^b - \vec{E}_i^b \wedge \vec{H}^a] \cdot \vec{n}\, dS + \int_S [\vec{E}^a \wedge \vec{H}_r^b - \vec{E}_r^b \wedge \vec{H}^a] \cdot \vec{n}\, dS \\ &= I'(S) + I''(S) \end{aligned} \tag{11}$$

The values of the incident field are introduced, and $I'(S)$ is calculated immediately:

$$I'(S) = \frac{4\pi}{i\eta k} \vec{E}_0 \cdot \vec{F}^a(\vec{n}_i) \tag{12}$$

In addition, it is easy to show that $I''(S) = 0$. The theorem of reciprocity (Part 1, page 16) can be applied to domain Ω for the field quantities $(\vec{E}^a, \vec{H}^a)$ and $(\vec{E}_r^b, \vec{H}_r^b)$. This gives:

$$\int_{S \cup S_\infty} (\vec{E}^a \wedge \vec{H}_r^b - \vec{E}_r^b \wedge \vec{H}^a) \cdot \vec{n}\, d\sigma = 0 \tag{13}$$

But on S_∞ the fields considered satisfy the radiation condition, whatever the state. The contribution of S_∞ is therefore zero. That of S results from it, so:

$$I(S) = \frac{4\pi}{i\eta k} \vec{E}_0 \cdot \vec{F}^a(\vec{u}_i) \tag{14}$$

This result is true whether S is outside S_a or identical with it.

(b) Breakdown of S_a

It is helpful to distinguish two zones on S_a. One, called S'_a, is such that at each point, whatever the state, the fields obey a relationship of the type:

$$\vec{E}^{a,b} = \zeta \vec{H}^{a,b} \wedge \vec{n} \tag{15}$$

where ζ is a surface impedance. This is the case for a perfect conductor $\zeta = 0$ or for a metal of finite conductivity. The contribution of S'_a to the calculation of $I(S_a)$ is, therefore, zero, as can be verified. The other zone of S_a is called S''_a and corresponds to the point at which a transmission generator or receiver load is connected. In this zone, the behaviour of the field can be described in a relatively simple way, and related to the quantities describing the antenna operation, taking account of its mode of connection. Two examples are given below.

• For a wire antenna, S''_a can be assimilated to a straight cylinder with generating lines parallel to Oz (Figure 19), of height h and radius ρ. Whatever the mode of operation, a distribution of $\vec{E}$, $\vec{M}$ around the origin has associated with it the voltage V and the current I defined by:

$$V^{a,b} = \int_{-(h/2)}^{+(h/2)} E^{a,b} \cdot e_z \, dz; \quad (\rho = 0) \tag{16}$$

$$I^{a,b} = \int_0^{2\pi} H^{a,b} \cdot e_\varphi \rho \, d\varphi; \quad \left(\rho = a, \quad |z| = \frac{h}{2}\right) \tag{17}$$

This type of definition is possible if the cut is sufficiently small. In transmission mode, the field in S''_a is assumed to be of the form:

$$\vec{E}^a = E(z) \, \vec{e}_z \tag{18}$$

$$\vec{H}^a = H(\varphi) \, \vec{e}_\varphi \tag{19}$$

Relating this to the integral, one obtains the following:

$$I(S_a) = I^a V^b - I^b V^a \tag{20}$$

• For an antenna with apertures fed by a guiding structure, the calculation is carried out differently. For S''_a (Figure 20) a plane along the waveguide axis is

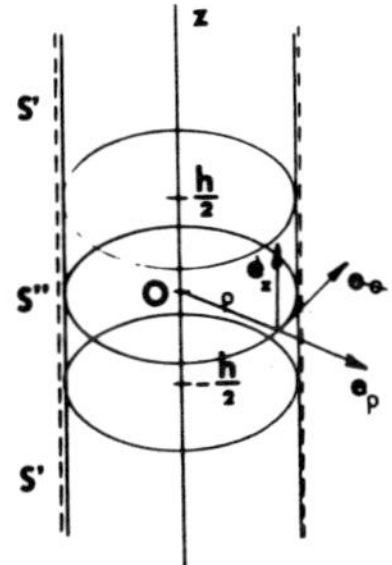

Fig. 19. *Excitation zone of a wire antenna.*

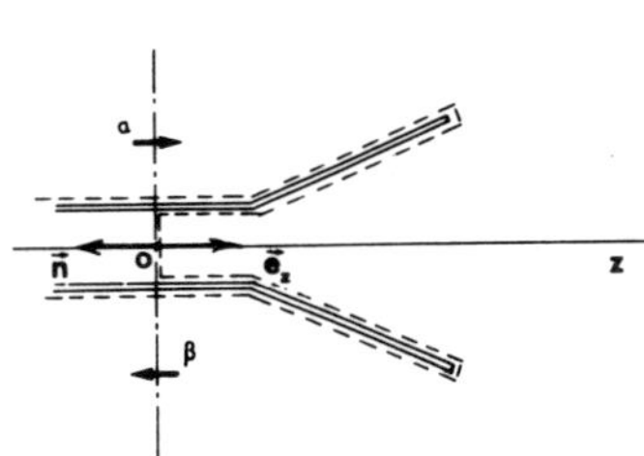

Fig. 20. *Excitation zone of a radiating aperture.*

chosen and is assumed to be monomodal and situated at a sufficient distance from the aperture for all the evanescent modes to be considered negligible. At any point of S''_a, and whatever the state, the tangential field is of the form:

$$\vec{E}_t^{a,b} = (\alpha^{a,b} + \beta^{a,b})\,\vec{e}_0 \tag{21}$$

$$\vec{H}^{a,b} = (\alpha^{a,b} - \beta^{a,b})\,\vec{h}_0 \tag{22}$$

with

$$\vec{e}_0 = \eta_{g_0}\vec{h}_0 \wedge \vec{e}_z$$

where $\vec{e}_0$ and $\vec{h}_0$ are the specific vectors associated with the mode being propagated, η_{g_0} its wave impedance and α, β coefficients proportional to the amplitude of the incident and reflected waves. The calculation gives:

$$I(S_a) = \frac{2}{\eta_{g_0}}(\alpha^a\beta^b - \alpha^b\beta^a)\iint_{S''_a} e_0'^2\, d\sigma \tag{23}$$

where e'_0 is the norm of $\vec{e}_0$.

3: Determining the equivalent generator

By equating the values of $I(S_a)$ calculated previously, for example for wire antennas:

$$I^aV^b - I^bV^a = \frac{4\pi}{i\eta k}\,\vec{E}_0\cdot\vec{F}^a(\vec{u}_i) \tag{24}$$

Let:

$$\mathscr{E}'^b = \frac{4\pi}{i\eta k}\,\vec{E}_0\cdot\frac{\vec{F}^a(\vec{u}_i)}{I^a} \tag{25}$$

The input impedance of antenna Z_e can be introduced, such that $V^a = Z_eI^a$. It follows that:

$$V^b = \mathscr{E}'^b + Z_eI^b \tag{26}$$

The interpretation is straightforward, taking account of the direction of orientation chosen for the current I^b (Figure 21). In reception mode, the antenna behaves in relation to its charge as a generator of electromotive force $\mathscr{E}'^b$ and of internal impedance equal to its input impedance in transmission mode. The electromotive force depends both on the incident wave and on the antenna because of its radiation characteristic in transmission mode. Note that the current cancels due to the terms $\vec{F}^a/I^a$.

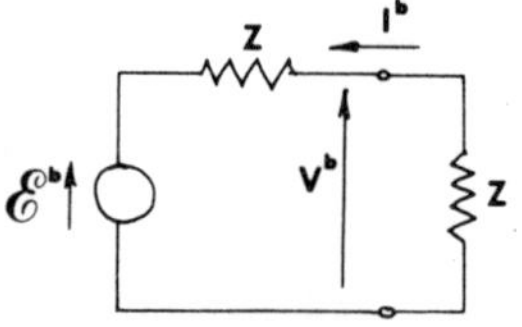

Fig. 21. *Equivalent circuit of an antenna functioning in reception mode.*

If the antenna is subjected to an angular spectrum of plane waves $\vec{E}_0(\vec{u})$, the electromotive force is found by integration:

$$\mathscr{E}'^b = \frac{4\pi}{i\eta k} \int \vec{E}_0(u) \cdot \vec{F}_r^a(\vec{u})\, d\bar{\omega}(\vec{u}) \tag{27}$$

where $d\bar{\omega}$ is the solid angle element in direction $\vec{u}$ and $\vec{F}_r^a$ is the reduced characteristic $\vec{F}^a/\vec{I}^a$.

C: Antenna coupling

1: Matrix representation

In the previous section the behaviour of the antenna in a given field was considered, without the source creating the field being taken into account. A more complete study involves an analysis of the coupling between the transmitting antenna and the receiving antenna. To achieve this, it is helpful to associate with them and with the medium separating them, a linear quadrupole. This quadrupole can be characterized in various ways, depending on the values adopted for the description of the antenna operation. For wire antennas, an impedance—or admittance—matrix is introduced, relating the currents and voltages at the antenna terminals. For two antennas (Figure 22):

$$V_1 = Z_{11}I_1 + Z_{12}I_2 \tag{28}$$

$$V_2 = Z_{21}I_1 + Z_{22}I_2 \tag{29}$$

where Z_{ii} represents the input impedance of the ith antenna in the presence of the other, with Z_{12} and Z_{21} as the coupling impedances. These impedances depend on the characteristics of each of the antennas, on their relative positions, the radiation medium and, possibly, on the circuit elements between the antennas and the terminals of the quadrupole. When the radiation medium and elements are reciprocal, $Z_{12} = Z_{21}$, as will be seen below.

When the antennas are at a long distance from one another—as is often the case in telecommunication problems—the coupling is sufficiently weak to allow it to be ignored for the transmitting antenna. For the ith transmitting antenna the following will be found:

$$V_i = Z_{ii}I_i + Z_{ij}I_j \simeq Z_{ii}I_i = Z_iI_i \tag{30}$$

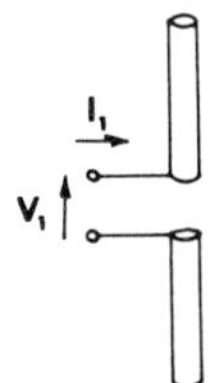

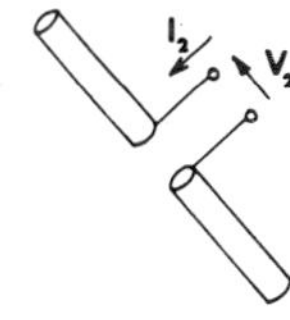

Fig. 22. *Coupling between two wire antennas.*

where Z_i is the input impedance of this antenna when it is alone in the space. For the reception antenna, the two terms can be comparable and it is not possible *a priori* to neglect one in preference for the other. The impedance Z_{ii} will be replaced by the input impedance for transmission. The voltage of the ith antenna at reception will be:

$$V_i \simeq Z_i I_i + Z_{ij} I_j \tag{31}$$

Equating this expression with the one obtained previously (26) allows the electromotive force of the equivalent generator to be simply expressed as a function of the coupling impedance:

$$E_i' = Z_{ij} I_j \tag{32}$$

The total current flowing in the load Z_{ic} is therefore:

$$I_i = \frac{E_i'}{Z_i + Z_{ic}} \tag{33}$$

For aperture antennas, supplied by guiding structures, the use of a scattering matrix is preferable. The unit formed by the two antennas and the medium that separates them behaves, for each of the structures, as a junction. Denoting by α_i and β_i, respectively, the amplitudes of the incident and reflected waves in a suitably chosen reference plane P_i,

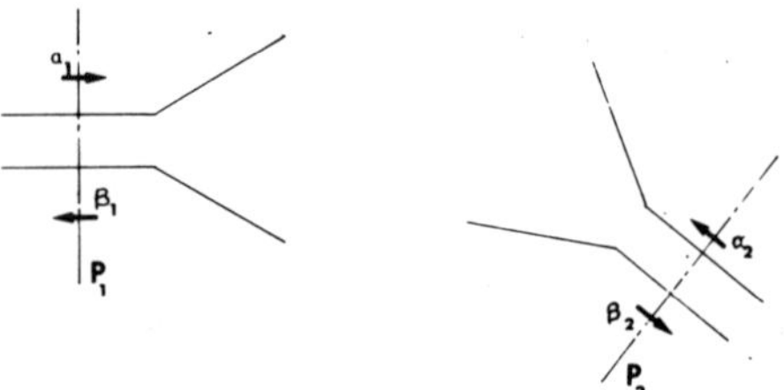

Fig. 23. *Coupling between aperture antennas.*

the following can be written:

$$\beta_1 = S_{11}\alpha_1 + S_{12}\alpha_2 \tag{34}$$

$$\beta_2 = S_{21}\alpha_1 + S_{22}\alpha_2 \tag{35}$$

where the coefficients S_{ii} and S_{ij} represent coefficients of reflection and transmission (or coupling). If the coupling is weak, S_{ii} can be replaced by S_i, the value of the reflection coefficient for the ith antenna transmitting alone. With respect to the load, characterized by its reflection characteristic ρ_i referred to the plane P_i, the antenna behaves as a voltage generator, the reflection coefficient of which, with respect to the same reference plane, would be S_i. For the ith reception antenna, it is easy to obtain:

$$\beta_i = \frac{S_{ij}\alpha_j}{1 - \rho_i S_i} \tag{36}$$

Note 1:
Whatever the mode of representation chosen, the electromotive force of the equivalent generator is proportional to the coupling parameter.

Note 2:
The above considerations can easily be extended to the case of N coupled antennas.

2: Calculation of the coupling parameters

The coupling parameters can be calculated using the theory of reciprocity applied to a domain Ω with boundary $\Sigma = S_1 \cup S_2 \cup S_\infty$ where S_1 and S_2 are two surfaces surrounding the antennas in question, and S_∞ is a sphere of large radius, surrounding S_1 and S_2 (Figure 24). The properties specified for Ω were listed in section 1 above. The two states considered are defined in the following way:

State a: Antenna 1 functions in transmission, antenna 2 in reception. The fields at each point of Ω are denoted by $\vec{E}^a$, $\vec{H}^a$.

State b: Antenna 1 functions in reception, antenna 2 in transmission. The fields at each point of Ω are denoted by $\vec{E}^b$, $\vec{H}^b$.

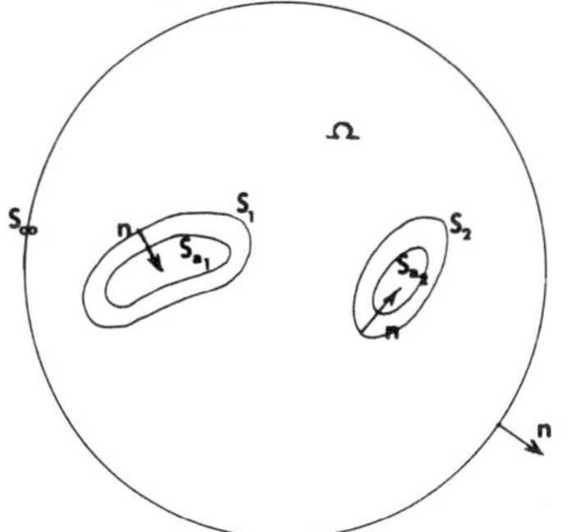

Fig. 24.

The theorem of reciprocity allows the following to be written:

$$\int_{S_1 \cup S_2} (\vec{E}^a \wedge \vec{H}^b - \vec{E}^b \wedge \vec{H}^a) \cdot \vec{n}\, dS = 0 \tag{37}$$

where u is the unit vector normal to $S_1 \cup S_2$ directed out of Ω. The contribution of S_∞ disappears, since in both the states considered the fields satisfy the radiation conditions.

Important note
It should be specified that during the change from transmission to reception, the generator must be replaced by a load with impedance equal to the internal impedance of the generator.

The choice of surface S_i is arbitrary provided it surrounds the ith antenna. In the specific case of $S_i = S_{ai}$, then (20) and (37) for wire antennas give:

$$I_1^b V_1^a - I_1^a V_1^b = -(I_2^b V_2^a - I_2^a V_2^b) \tag{38}$$

If, in reception mode, the antennas are supplied by ideal voltage generators—which implies that in passive mode the antennas are short-circuited—this reduces to:

$$I_{1cc}^b V_1^a = I_{2cc}^a V_2^b \tag{39}$$

If, correspondingly, ideal current generators are used, then:

$$I_1^a V_{1co}^b = I_2^b V_{2co}^a \tag{40}$$

By definition, the coupling impedance is:

$$Z_{ij} = \left.\frac{V_i}{I_j}\right|_{I_i=0} = \frac{V_{ico}}{I_j} \tag{41}$$

As a result of the previous relationships, the following would be obtained:

$$Z_{ij} = \frac{V_{ico}}{I_j} = \frac{V_{ico}}{I_i} = Z_{ji} \tag{42}$$

But any other surface can be chosen for S_i. It is interesting to note that one of these is made up of a hemisphere, closed by a plane Π (Figure 25). If the radius of the spherical part tends towards infinity, only the integral on plane Π will remain. Surrounding antenna 2 by such a surface, for example, gives:

$$I_1^a V_{ico}^b = -\int_\Pi (\vec{E}^a \wedge \vec{H}^b - \vec{E}^b \wedge \vec{H}^a) \cdot \vec{n}\, dS = -I(\Pi) \tag{43}$$

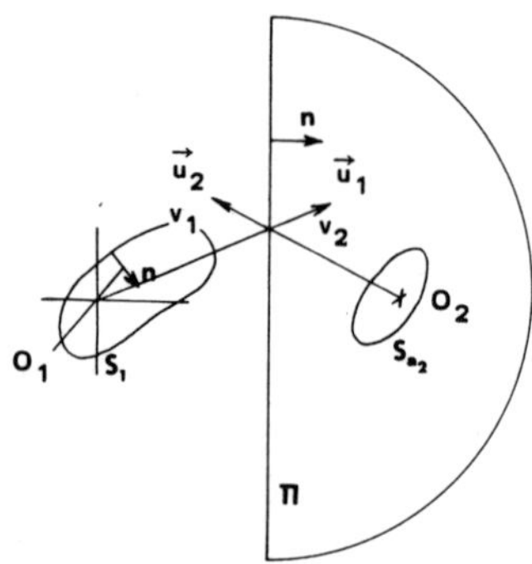

Fig. 25.

This expression allows the coupling impedance to be calculated from the field values on π:

$$Z_{12} = \frac{V_{1co}^b}{I_2^b} \cdot \frac{I_1^a}{I_1^a} = -\frac{1}{I_1^a I_2^b} I(\Pi) \tag{44}$$

Considering, now, two antennas supplied by identical waveguide structures, the choice of $S_i = S_{ai}$ means that from (23) and (37) the following can be obtained:

$$\alpha_1^a \beta_1^b - \alpha_1^b \beta_1^a = -(\alpha_2^a \beta_2^b - \alpha_2^b \beta_2^a) \tag{45}$$

For such antennas, it is interesting to consider the case of matched generators. This gives:

$$\alpha_{1a}^a \beta_{1a}^b = \alpha_2^b \beta_{2a}^a \tag{46}$$

from which the coupling coefficient is deduced:

$$S_{12} = S_{21} = \frac{\alpha_1^a \beta_{1a}^b}{\alpha_1^a \alpha_2^b} = -\frac{1}{\alpha_1^a \alpha_2^b} I(\Pi) \tag{47}$$

The coupling parameters can therefore be obtained simply, once the distributions $\vec{E}^a$, $\vec{H}^a$ and $\vec{E}^b$, $\vec{H}^b$ are known on a surface surrounding one of the antennas and, in particular, on a plane that separates them. The strict determination of these fields is a difficult diffraction problem: it involves calculating the field radiated by the transmitting antennas. If, however, the antennas are at a great distance from each other, taking account of the attenuation of the radiated fields, it can be assumed as a preliminary approximation that in the plane Π, situated at a comparable distance from each of the antennas, the fields $\vec{E}^a$, $\vec{H}^a$ and $\vec{E}^b$, $\vec{H}^b$ are respectively identical with the fields radiated by antennas 1 and 2, if they were alone in the space. The error goes as d^{-2}, where d is the distance between antennas. The far-field expressions could be used. By introducing the transmission radiation characteristics of each of the antennas, the following is found:

$$I(\Pi) = \int_\Pi \vec{F}_1^a \vec{F}_2^b (\vec{u}_2 - \vec{u}_1) \cdot \vec{n} + (\vec{F}_2^b \cdot \vec{u}_1)(\vec{F}_1^a \cdot \vec{n}) - (\vec{F}_1^a \cdot \vec{u}_2)(\vec{F}_2^b \cdot \vec{n}) \frac{e^{-ik(r_1 + r_2)}}{\eta r_1 r_2} \, dS \tag{48}$$

where the meaning of r_1, r_2, $\vec{u}_1$, $\vec{u}_2$, $\vec{n}$ is given in Figure 25. When the antennas are close together, it may be helpful to use the representation of the fields in the form of an angular spectrum of plane waves. The integral can be calculated approximately using the stationary phase method.

3: Study of a specific case

This concerns two directive antennas, more or less pointed at each other, such that, in the calculation of $I(\Pi)$, it can be assumed that:

$$\left.\begin{aligned} \vec{u}_2 = -\vec{u}_1 = -\vec{n} \\ \vec{F}_1 \cdot \vec{n} = \vec{F}_2 \cdot \vec{n} = 0 \end{aligned}\right\} \tag{49}$$

The formulae given above can be simplified greatly, leaving:

$$I(\Pi) = \frac{2}{\eta} \int \vec{F}_1^a(\vec{u}) \cdot \vec{F}_2^b(\vec{u})\, d\bar{\omega}(\vec{u}) \tag{50}$$

These formulae are commonly used in practice. Coupling between one antenna and the other is maximal when the electric field distributions they create in the plane Π are identical.

D: Power balance

1: General formula

The discussion will be based on the case of a wire antenna exposed to a plane wave.

The power transported by the incident wave can be characterized by Poynting's vector:

$$\vec{P}_i = \frac{1}{2}(\vec{E}_i \wedge \vec{H}_i^*) = \frac{1}{2\eta} E_0'^2 \vec{u}_i \tag{51}$$

The active power intercepted by the antenna is expressed as:

$$W' = \frac{1}{2} \operatorname{Re} \left\{ \iint_S (\vec{E} \wedge \vec{H}^*) \cdot \vec{n}\, dS \right\} \tag{52}$$

where $\vec{E}$ and $\vec{H}$ are the total fields in reception mode. This power is distributed between power dissipated in the antenna and active power supplied to the load:

$$W' = W_c' + W_d' \tag{53}$$

Any losses will be ignored in the following. The general formula (52) allows the calculation of power to be carried out from the field expressions. The discussion is simpler when related to the equivalent circuit of the antenna, functioning in the reception mode.

The *reflection coefficient* may be introduced:

$$\rho_c = 1 - \left| \frac{Z^* - Z_c}{Z + Z_c} \right|^2 \tag{54}$$

$E_0'^2$ can be expressed as a function of Poynting's incident vector, the input resistance and the gain (Chapter 4, equation (29)).

$$F_a'^2(\vec{u}_i) = \frac{\pi R G_a(\vec{u}_i)}{\eta} \tag{55}$$

Let:

$$\left.\begin{aligned} \vec{e}_0 &= \vec{E}_0 / E_0' \\ \vec{f}_0 &= \vec{F}_a / F_a' \end{aligned}\right\} \tag{56}$$

The polarization factor is defined by:

$$\rho_p(\vec{u}_i) = |\vec{e}_0 \cdot \vec{f}_0|^2 \tag{57}$$

The active power supplied to the load is, therefore:

$$W'_c = \operatorname{Re}(W_a) = \frac{\lambda^2}{4\pi} P'_i G(\vec{u}_i)\rho_p(\vec{u}_i)\rho_c \tag{58}$$

The active re-radiated power can be calculated from the formula:

$$W'_r = \frac{1}{2} \operatorname{Re}\left\{\iint_S (\vec{E}_r \wedge \vec{H}_r^*)\cdot\vec{n}\, dS\right\} \tag{59}$$

where $\vec{E}_r$ and $\vec{H}_r$ were defined in (6) and (7).

Formula (58) is fundamental. It allows the influence of the various parameters to be revealed.

2: Influence of the load

This is basically a problem concerning the power delivered by a generator as a function of its load impedance. The load is matched to the generator when $z_c = z^*$; in this situation the charge factor is 1. The power extracted by the load is therefore a maximum at $E'^2/8R_c$. For a more complete discussion, the complex impedance plane $(R/R_c,\ X + X_c/R_c)$ can be studied (Figure 26) to reveal the location of the points with constant load factor: a plot of these forms a set of circles with focus $R/R_c = 1$. If the ratio R/R_c is fixed, it may be of interest to cancel out the total reactance to give the largest charge factor. This will always be between 0 and 1.

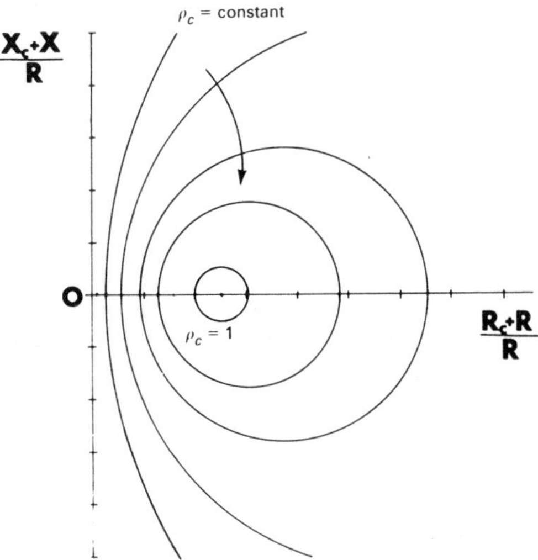

Fig. 26. *Variations in the load factor in the plane of complex impedances.*

3: Influence of polarization

The polarization factor ρ_p is the square of the scalar product modulus of two normalized vectors: it is therefore between 0 and 1. It depends only on the polarization of the incident field—assumed here to be constant for any antenna—and on the transmission characteristic in direction $\vec{u}_i$. Each of these complex vectors can be resolved into two orthogonal vectors $\vec{e}_1$, $\vec{e}_2$, situated in any plane normal to $\vec{u}_i$. This gives:

$$\vec{e}_0 = \cos\alpha\,\vec{e}_1 + i\sin\alpha\,\vec{e}_2 \qquad 0 \leqslant \alpha \leqslant \frac{\pi}{4}$$
$$\vec{f}_0 = \cos\beta\, e^{i\theta_1}\,\vec{e}_1 + i\sin\beta\, e^{i\theta_2}\,\vec{e}_2 \qquad (60)$$

The axes $\vec{e}_1$, $\vec{e}_2$ can be chosen so that the ellipse of $\vec{e}_0$ intersects $\vec{u}_i$ at right angles $\vec{u}_i$. The value of the polarization factor can be deduced from this:

$$\rho_p = \frac{1}{2}[1 + \cos 2\alpha \cos 2\beta - \sin 2\alpha \sin 2\beta \cos(\theta_1 - \theta_2)] \qquad (61)$$

Figure 27 gives a number of particular configurations. It is important to note the conditions in which the polarization coefficient takes its extreme values:

- ρ_p takes the value 1 when the two ellipses (e_0) and (f_0) are identical and oriented in opposite directions with respect to a normal to their plane.
- ρ_p takes the value 0 when the two ellipses are equal, orthogonally arranged and oriented in the same direction.

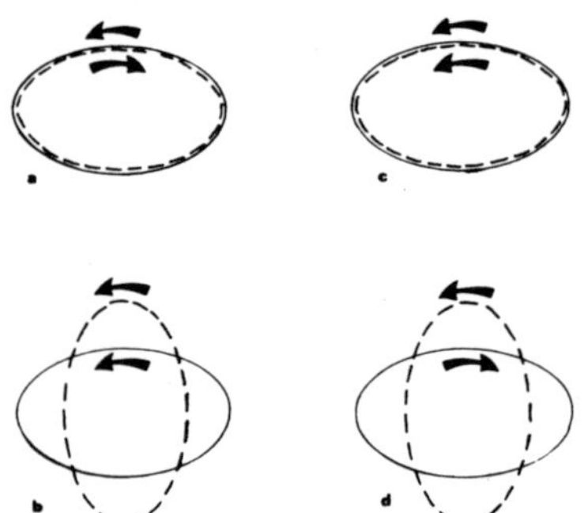

Fig. 27. *Polarization ellipses associated with* l_0 (—) and f_0 (– – –); (*a*) $\rho_p = 1$; (*c*) $\rho_p = \cos^2 2a$; (*b*) ρ_p; (*d*) $\rho_p = \sin^2 2a$.

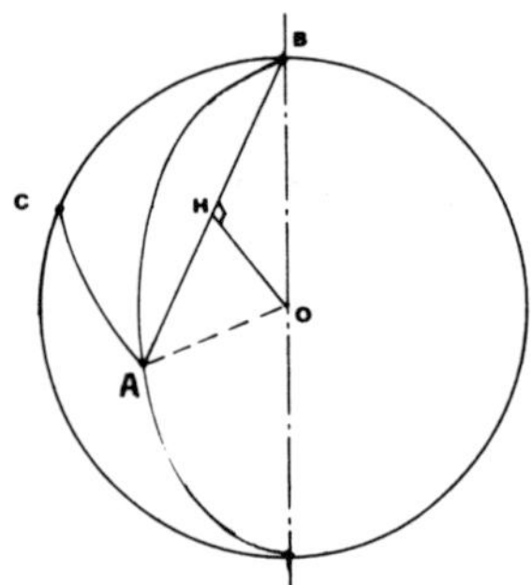

Fig. 28. *Graphic construction allowing the polarization factor to be determined.*

The spherical representation (Figure 28) makes the determination of the polarization factor immediately clear. On a great circle with unit radius, the arc $BC = 2\alpha$ is plotted; then, C, the great circle inclined by $\theta_1 - \theta_2$ on the extension of BC. On the second great circle, $\overline{CA} = 2\beta$ is plotted. This gives a spherical triangle ABC such that $\rho_p = \overline{OH}^2$.

4: Directive properties of receive antennas

The power W'_c supplied to the load depends on the direction of the incident field through the intermediary of the transmission gain $G(\vec{u}_i)$ and the polarization factor $\rho_p(\vec{u}_i)$. When, for each direction u_i, the polarization is matched, then W'_c varies as $G(\vec{u}_i)$. The antenna is also said to have the same directive properties in both transmission and reception modes. The characteristic radiation and reception surface coincide—to the nearest constant factor—as long as the polarizations are matched.

It is helpful to note here that the re-radiation occurs in a manner different from that of the transmission radiation.

E: Equivalent areas

1: Definition

The powers W'_c and W'_r are proportional to the norm of the Poynting vector P'_i associated with the incident wave. The proportionality is expressed using a quantity with the dimensions of area. S_c and S_r are defined, such that:

$$W'_c = S_c P'_i \tag{62}$$

$$W'_r = S_r P'_i \tag{63}$$

S_c and S_r are known, respectively, as the absorption area and the diffusion area. From relationship (62), the following expression can be deduced straight away:

$$S_c(\vec{u}_i) = \frac{\lambda^2}{4\pi} G(\vec{u}_i)\rho_c\rho_p(\vec{u}_i) \tag{64}$$

Under the optimal conditions for antenna use $\rho_c = \rho_p = 1$, when:

$$S_{c_{\text{opt.}}}(\vec{u}_i) = \frac{\lambda^2}{4\pi} G(\vec{u}_i) \tag{65}$$

This clearly gives:

$$S_c(\vec{u}_i) = \rho_c\rho_p(\vec{u}_i) S_{c_{\text{opt.}}}(\vec{u}_i) \tag{66}$$

The importance of introducing the idea of area is that it makes the calculation of power easier, by clearly separating what arises from the antenna itself from what arises from the incident field. The latter is simply calculated from transmission mode formulae.

The absorption area has a simple physical meaning in the case of aperture antennas. For such antennas, and in the situation where the dimensions of the aperture are large compared to the wavelength, equating (65) and (67) of Chapter 5 gives the following result:

$$S_{c_{\text{opt.}}} = S \tag{67}$$

The optimal absorption area is equal to the aperture surface. This means that, in the aperture, the field is practically the same as would exist in the absence of the reception antenna.

For small or wire antennas, the absorption area must never be assimilated into the effective area of the antenna projected onto a plane perpendicular to the direction.

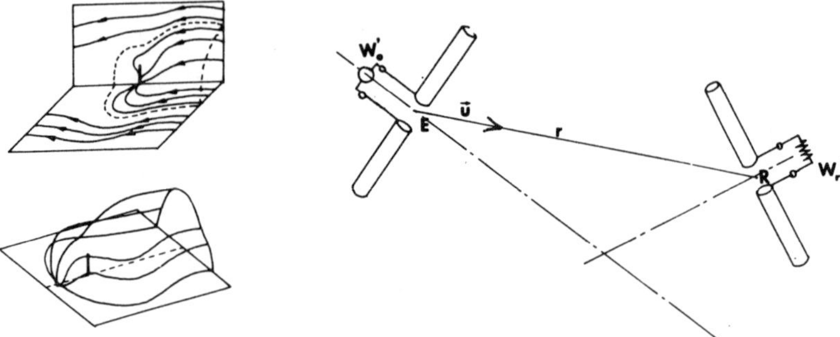

Fig. 29. *Definition of the absorption area for a linear antenna (Landstorfer and Meinke).*

Fig. 30. *Link between two antennas in free space.*

For a dipole, for example, the absorption area coincides with the surface inside which the lines of power flow to the antenna. This surface is, fortunately, larger than that of the antenna proper (Figure 29).

2: Applications

(a) Radio links in free space

A link is established between a transmitting antenna (E) and a reception antenna (R), with respective gains $G_e(\vec{u})$, $G_r(\vec{u})$. If a power W'_e is supplied to the transmitting antenna—assumed to be of unit efficiency loss—what is the power W'_r received by the antenna R? (Figure 30). For the purposes of this discussion, the coupling is assumed to be weak.

Calculation takes place in two stages. Antenna E creates around antenna R, located at $(r, \vec{u})$, a field the norm of which is:

$$E'^2 = \frac{\eta}{2\pi} \frac{W'_e G_e(\vec{u})}{r^2} \tag{68}$$

The Poynting vector associated with this field is:

$$P'_i = \frac{E'^2}{2\eta} = \frac{W'_e G_e(\vec{u})}{4\pi r^2} \tag{69}$$

From this, the power supplied to the load of antenna R is deduced:

$$W'_r = P'_i S_c = \frac{W'_e}{4\pi r^2} G_e(\vec{u})\rho_p(-\vec{u})\rho_c S_{\text{opt.}}(-\vec{u}) \tag{70}$$

or:

$$W'_r = \frac{\lambda^2}{(4\pi r)^2} G_e(\vec{u})G_r(-\vec{u})\rho_p(-\vec{u})\rho_c W'_e \tag{71}$$

The ratio W'_r/W'_e defines the attenuation of the waves in free space. In some simple cases, this attenuation can be corrected to take into account possible irregularities in the propagation medium (earth, ionosphere, various obstacles, etc).

(b) Radar equations

An antenna E transmits a power W'_e. A second antenna, R, receives the power W'_r re-radiated by a target, located at $\vec{r}_e$, $\vec{u}_e$ and $\vec{r}_r$, $\vec{u}_r$ from each antenna (Figure 31). Poynting's vector incident to the target is, as in the previous case:

$$P'_i = \frac{W'_e G_e(\vec{u})}{4\pi r_e^2} \tag{72}$$

Let $S_r(\vec{u}, \vec{v})$ be the echoing area of the target. As will be noted, this area depends on the direction of incidence $\vec{u}$ and the direction of re-radiation $\vec{v}$. Poynting's vector re-radiated at the level of antenna R is written:

$$P' = \frac{1}{4\pi r_r^2} S_r(\vec{u}_e, \vec{u}_r)P'_i \tag{73}$$

The received power W'_r can be deduced:

$$W'_r = P'S_c = \frac{\lambda^2}{(4\pi)^3 r_e^2 r_r^2} G_e(\vec{u}_e)G_r(\vec{u}_r)\rho_c\rho_p(\vec{u}_r)S_r(\vec{u}_e, \vec{u}_r)W'_e \tag{74}$$

The antenna can be matched to its load (circuit problem). On the other hand, the polarization factor cannot be controlled since there is generally depolarization of the wave re-radiated by the target.

F: Behaviour of an antenna in the presence of noise

1: Noise

It is important to characterize the behaviour of the antenna in relation to noise, in the same way as for the elements of a receiver chain. The antenna occupies a critical position, since it is involved in the capture of the signal where it is weakest.

Superimposed on the effective radiation there is usually a parasitic radiation arising from natural or artificial sources, which may be coherent or incoherent, polarized or unpolarized. Only thermodynamic radiation from the bodies surrounding the antenna is considered here. This thermodynamic radiation is incoherent and unpolarized. It is unavoidable.

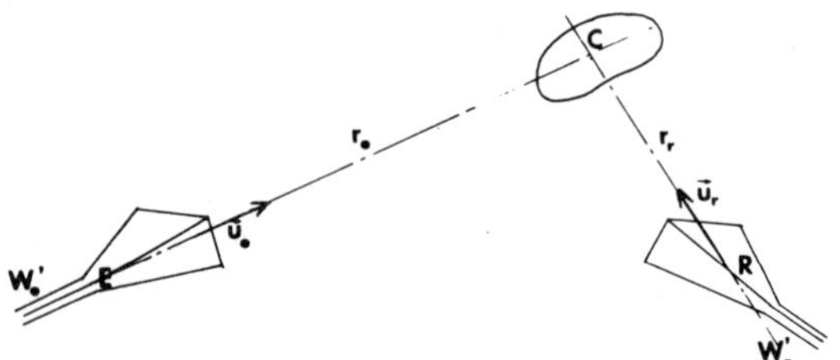

Fig. 31. *Configuration relating to the radar equation.*

• Summary of thermodynamic radiation of bodies:

Consider a body in thermodynamic equilibrium at temperature T. If this body is supplied with a power W, it absorbs a fraction aW, with $0 \leqslant a \leqslant 1$. Here a is the absorption constant of the body, and is 1 for a black body. This body re-emits, in the form of incoherent and unpolarized electromagnetic radiation, the power it has absorbed. An element dS, with normal $\vec{n}$, radiates in the solid angle $d\bar{\omega}$ about the direction θ, measured from the normal, a power in the frequency band Δf:

$$dW_e = aE(f, T) \cos\theta \, d\bar{\omega} \, dS \, \Delta f$$

where $E(f, T)$ is the spectral density given by Planck's formula.

$$E(f, T) = \frac{2hf^3}{c^2} \frac{1}{\exp(hf/kT) - 1}$$

where h is Planck's constant and k is Boltzmann's constant.

In radio technology, Rayleigh's approximation is often sufficient, and it is valid if $hf/kT \ll 1$:

$$E(f, T) \simeq \frac{2kT}{\lambda^2}$$

To fix the orders of magnitude,

$$\frac{hf}{kT} = 4.75 \times 10^{-11} \frac{f}{T},$$

where f is in Hz and T in kelvins.

2: Noise temperature

It is assumed that the antenna exchanges thermal energy with an enclosure Σ formed by a black body in equilibrium at the temperature T. A surface element $\Delta\Sigma$ of this enclosure transmits, in a solid angle $\Delta\omega$ about direction θ of the antenna, relative to $\vec{n}$, a power in the band Δf equal to:

$$\Delta W = E \cos\theta \, \Delta\Sigma \, \Delta\bar{\omega} \, \Delta f \tag{75}$$

Consider more specifically $\Delta\bar{\omega}$, the solid angle under which is the capture area for the antenna with midpoint M of $\Delta\Sigma$:

$$\Delta\bar{\omega} = \frac{S_c(\vec{u})}{r^2} \tag{76}$$

where r is the distance separating the antenna from the element of $\Delta\Sigma$.

Since the radiation is unpolarized, it can be resolved into two orthogonal polarizations, one of which is matched to the antenna. The power is distributed equally between these two polarizations, which leads to a polarization factor equal to $\frac{1}{2}$, so:

$$S_c(\vec{u}) = \frac{1}{2}\frac{\lambda^2}{4\pi} G(\vec{u}) \tag{77}$$

where $G(\vec{u})$ is the antenna gain in direction $\vec{u}$, and where the load is assumed to be matched to the antenna. By introducing the solid angle $\Delta\bar{\omega}'$ which $\Delta\Sigma$ of the antenna subtends, it is easy to establish that the noise power captured by the antenna is:

$$\Delta W_a = \frac{kT}{4\pi} G(\vec{u})\, \Delta\bar{\omega}'\, \Delta f \tag{78}$$

The total power is obtained by integration over the whole enclosure:

$$W_a = \frac{kT}{4\pi} \Delta f \int G(\vec{u})\, d\omega'(\vec{u}) = kT\, \Delta f \tag{79}$$

This formula is in the same form as that allowing the thermal noise in the resistances (Johnson effect) to be obtained. The equivalent circuit of the reception antenna can, therefore, be completed by adding a noise generator capable of supplying the available power:

$$W_a = \frac{\overline{e_b^2}}{4R} = kT\, \Delta f \tag{80}$$

In practice, the antenna is surrounded by bodies at different temperatures (ground, space objects, etc), such that the total noise power in the band Δf is expressed:

$$W_a = \frac{k}{4\pi} \Delta f \int T(\vec{u}) G(\vec{u})\, d\bar{\omega}'(\vec{u}) = kT_A\, \Delta f \tag{81}$$

where T_A is the antenna noise temperature:

$$T_A = \frac{1}{4\pi} \int T(\vec{u}) G(\vec{u})\, d\bar{\omega}'(\vec{u}) \tag{82}$$

It will be found that $T_A = T$ in the case of a uniform distribution of temperature. Note that the noise temperature does not constitute specific data concerning the antenna. It allows the behaviour of the antenna to be characterized in a given external environment: in direction $\vec{u}$, the temperature $T(\vec{u})$ is balanced by the gain $G(\vec{u})$. For an antenna installed on the ground, the noise power arises partly from outer space (temperature of the order of some tens of kelvins), partly from the ground (noise temperature of the order of 300 K).

If the antenna is highly directive and pointed towards the zenith, it will not be sensitive to radiation from earth and its noise temperature will be similar to that of outer space. If the antenna is misaligned and as a result intercepts some radiation from earth, its noise temperature will increase very rapidly.

For a non-directive antenna, the noise temperature hardly depends on its orientation at all, since it always captures about the same noise power from the ground.

3: Applications

The phenomenon of emission of thermal radiation is very general, and can be extended to all absorbant media situated between the source and the receptor. It can also apply to natural propagation media (ionosphere, atmosphere, etc), or else to the waveguide structure linking the antenna to the receiver. If α is the attenuation constant of the waves in one of these media, and the transmission constant a is the fraction of the power absorbed then:

$$a = 1 - e^{-2\alpha l} \tag{83}$$

If the available noise power at the input of the medium is $W_e = kT_e\,\Delta f$, the power available at the output W_s is such that:

$$W_S = (1 - a)kT_e\,\Delta f + akT\,\Delta f \tag{84}$$

where T is the temperature of the medium. Associated with the power W_s is the temperature:

$$T_{eq} = (1 - a)T_e + aT \tag{85}$$

If the losses are small

$$T_{eq} \simeq T_e + 2\alpha l(T - T_e) \tag{86}$$

It is therefore possible to calculate increasingly precisely the noise temperature at the input to the receiver. This can be added to the noise temperature of the receiver itself.

Bibliography to Chapter 7

Brown J. A generalized form of the aerial reciprocity theorem, *Proc. I.E.E.*, vol. **103c**, no. 4, April 1958, pp. 1–4.

Collin R. E., Zucker F. J. *Antenna theory*, McGraw-Hill, 1969, Chap. 4 by Collin R. E., pp. 93–137.

De Hoop A. T. The N-port receiving antenna and its equivalent electrical network, *Phillips Res. Repts.*, vol. **30**, 1975, pp. 302–315.

De Hoop A. T., De Jong G. Power reciprocity in antenna theory, *Proc. I.E.E.*, vol **121**, no. 10, Oct. 1974, pp. 1051–1056.

Robieux J. Lois générales de la liaison entre radiateurs d'ondes. Application aux ondes de surface et à la propagation. *Annales de Radioélectricité*, Part I, vol. **XIV**, no. 57, July 1959—pp. 186–229: Part II, vol. **XV**, no. 59, January 1960, pp. 28–77. Part III, vol. **XV**, no. 62, October 1960, pp. 305–330.

Chapter 8

Numerical methods

A: Introduction

The calculation of the radiated field from a known distribution of equivalent sources does not present any particular problems, even if, in some cases, the means of calculation that have to be used take a great deal of effort (calculation of surface integrals). It is also possible, in some cases, to calculate the input impedance of the antenna from this distribution. The values for the equivalent sources can be deduced from measurements (see Chapter 9), or *a priori* from physical considerations. In the latter case, there may be some difficulty with antennas of complex structure. In addition, the demand for equipment providing ever-improved performance standards requires considerable precision in calculations, incompatible with approximation.

To achieve this precision, one method consists of determining, by the most exact calculation possible, the distribution of equivalent sources. The integral formulation of radiation problems can help in this by representing the distribution of these sources, at a given point, very intuitively, as the result of a balance in contributions from applied sources and induced sources in the antenna. The balance equation (or more precisely, the expression of the boundary conditions) generally consists of integral or integro-differential linear equations. These can be reduced to linear equation systems of high rank, which can be solved by computer.

This procedure has become more popular with the developments in computer performance (both in rapidity and capacity). It is characterized by its rigour and generality: it can be applied, at least in principle, to any configurations. In practice, for large antennas, the calculation resources required are prohibitive.

Another approach can be envisaged for such large antennas. This uses the geometric theory of diffraction and thus provides a corrective term to the simpler methods based on geometric optics (see Part 1, page 95). With this method, it is not necessary to use equivalent sources as intermediaries in arriving at the radiated field. This is obtained at one point by superposition of the fields associated with various rays passing through that point (incident, reflected, refracted or diffracted). When the listing and calculation of these rays is possible, this procedure is very useful since it dispenses with the integration of equivalent sources over a large surface.

Only the integral formulation will be considered here. There is no question of discussing all the methods available for use, along with their many variants, in detail. The interested reader is referred to the bibliography at the end of this chapter. The object of this chapter is to show the mechanisms available for obtaining the equations to be solved and, with the use of simple examples, to illustrate the possibilities offered by these numerical methods, as well as the problems involved in their application in practice.

B: Integral formulation principle

As has already been seen, the radiated field at any point arises partly from known applied sources and partly from unknown induced sources. This second part is expressed with integrals relating to the induced sources. From the fact that the behaviour of the total field is locally imposed (by boundary conditions, passage conditions, etc), an integral relationship is obtained between the induced sources and the applied sources. This relationship can be used to calculate the induced sources.

1: Diffraction by a metallic body

(a) Maue's equation

Consider, firstly, the simplest case in terms of form, which consists of determining the currents induced on the surface of a perfectly conducting body (a reflector, for example) when it is subjected to an applied field $\vec{E}_a$, $\vec{H}_a$ (due, for

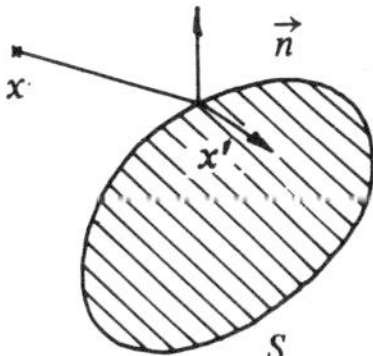

Fig. 32. *Diffraction by a conductive body.*

example, to a primary source). Let S (Figure 32) be the surface of the body, assumed to be perfectly conducting. The total field at any point external to S is:

$$U_S\vec{E}(x) = \vec{E}_a(x) - \frac{1}{i\omega\varepsilon}[k^2 + \text{grad div}] \int_S G(x; x')\vec{J}(x')\,dx' \tag{1}$$

$$U_S\vec{H}(x) = \vec{H}_a(x) - \text{curl} \int_S G(x, x')\vec{J}(x')\,dx' \tag{2}$$

with:

$$\vec{J}(x) = \vec{n} \wedge \vec{H}(x); \quad x \in S$$

Here $\vec{n}$ is the unit vector, normal to S and directed out of it; U_s is the characteristic function of the domain exterior to S.

The second equation allows $\vec{H}$ to be calculated as a function of its tangential component on S. The limiting case gives:

$$\lim_{x \to S} \left\{ \vec{n} \wedge U_S \vec{H}(x) + \vec{n} \wedge \operatorname{curl} \int_S G(x, x') \vec{J}(x')\, dx' \right\} = \vec{n} \wedge \vec{H}_a(x) \tag{3}$$

It can be shown that the first term is discontinuous across S (Part 1, Chapter 3, section 1). In the case under consideration, the following is obtained:

$$\frac{1}{2} \vec{J}(x) + \int_S \vec{n} \wedge [\vec{J}(x') \wedge \operatorname{grad}_x G(x, x')]\, dx' = \vec{n} \wedge \vec{H}_a(x); \quad x \in S \tag{4}$$

where the integral is to be taken in the sense of Cauchy's principal value. This is an integral equation of the second type in $\vec{J}$ (Maue's equation). Once $\vec{J}$ has been deduced, the transfer relationships in (1) and (2) allows the field to be calculated at any point in space. The formula obtained applies to any closed surface S, for any distribution of the applied field. It is assumed that the distribution of primary sources is not modified by the presence of the body. Note that it would also be possible to obtain an integral equation for $\vec{J}$ based on the integral representation of the electric field.

(b) Uniqueness of the solution

Before commencing the numerical solution of an integral equation, it is desirable to establish the uniqueness of the solution. For Maue's equation, which is a Fredholm equation of the second type, a complete discussion of the existence and uniqueness of the solution is possible (Fredholm's alternative). Only the most important results are given here.

Equation (4) always admits at least one solution, whatever the frequency. When, however, the adjoint homogeneous equation, associated with (4) and defined by:

$$\frac{1}{2} \vec{K}(x) - \int_S \operatorname{grad}_{x'} G^*(x, x') \wedge [\vec{n}' \wedge \vec{K}(x')]\, dx' = 0; \quad x \in S \tag{5}$$

admits a non-zero solution, then (4) admits an infinite number of solutions. This result may be surprising, since the problem under consideration admits only a single solution. The existence of spurious solutions can be explained as follows: in completing the vector product with $\vec{n}$ and taking the conjugate of the two members of (5), the following equation is found:

$$\frac{1}{2} \vec{L}(x) - \int_S \vec{n} \wedge [\operatorname{grad}_{x'} G(x, x') \wedge \vec{L}(x')]\, dx' = 0; \quad x \in S \tag{6}$$

with:

$$\vec{L} = \vec{n} \wedge \vec{K}^*$$

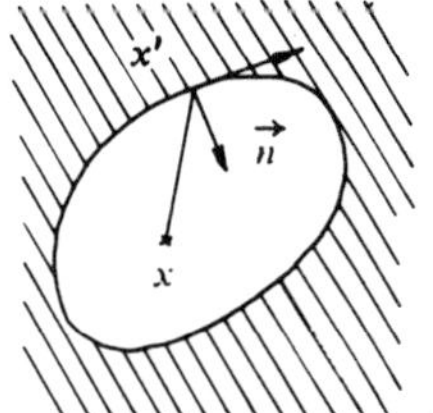

Fig. 33. *The equivalent internal case to the external problem of Figure 32.*

But this equation is simply that of Maue relative to the cavity inside S, without sources (Figure 33). Because of this, it admits non-zero solutions $\vec{L}_n$ for the discrete spectrum of resonance frequencies f_n in the cavity. The following important result follows:

For resonance frequencies in the cavity inside S, Maue's equation admits an infinite number of solutions.

For these frequencies, a solution to (4) can be written: $\vec{J} = \vec{J}_0 + \vec{J}_p$ where $\vec{J}_0$ is the physical solution to be found and $\vec{J}_p$ is a parasitic term linked to the cavity problem. In relation to J in (1) a field is found that can be broken down in the same way: $\vec{E} = \vec{E}_0 + \vec{E}_p$. The parasitic solutions can be easily distinguished from the physical solution by calculating the field inside S. By construction, $\vec{J}_0$ creates a zero field (perfectly conducting obstacle), while $\vec{J}_p$ gives a non-zero field. In order to be sure of the physical meaning of the solution $\vec{J}$, therefore, additional constraints are added to (4) to ensure that the field cancels inside S. Other techniques have been proposed to overcome this inconvenience in Maue's equation. This is a serious problem that should not be underestimated. Independently of any practical difficulty, it limits the domain of application of the integral formula to bodies in which the resonance spectrum inside the cavity is not too dense.

2: Radiation from an antenna

Another important case, in practical terms, is that in which the applied sources are situated inside a surface S, of which a part S', of whatever size, corresponds to a perfect conductor. Outside S the field is:

$$U_S\vec{E}(x) = -\frac{1}{i\omega\varepsilon}[k^2 + \text{grad div}]\int_S G(x, x')\vec{J}(x')\,dx' + \text{curl}\int_{S''} G(x, x')\vec{M}(x')\,dx' \tag{7}$$

where S'' is the part of S which is not conductive and therefore corresponds to an excitation zone of the antenna. A compatibility relation exists between $\vec{J}$ and $\vec{M}$, obtained by passing to the limit $x \to S$:

$$\frac{1}{2}\vec{M}(x) = \lim_{x \to S} \vec{n} \wedge \left\{-\frac{1}{i\omega\varepsilon}[k^2 + \text{grad div}]\int_S G(x, x')\vec{J}(x')\,dx' + \int_{S''} \text{curl}\, G(x, x')\vec{M}(x')\,dx'\right\} \tag{8}$$

This relationship, or other equivalent ones that can be deduced from it, allows the electrical density $\vec{J}$ to be calculated on the whole antenna from the single magnetic density $\vec{M}$ on the excitation aperture. ($\vec{J}$ on S' and M on S'' could also be determined from $\vec{J}$ on S''.) It can be used in a variety of ways, depending on the degree of rigour required and according to the configuration of the antenna around its zone of excitation.

(a) Wire antennas

1. Transmitting antennas This refers to antennas formed from wires of negligible diameter compared to their length. An important example is that of the linear antennas to be studied below (Figure 34). The generator will be modelled by a layer of longitudinal magnetic currents with revolutional symmetry, creating a known applied field:

$$\vec{E}_a(x) = \text{curl} \int_{S''} G(x, x')\vec{M}(x')\, dx' \tag{9}$$

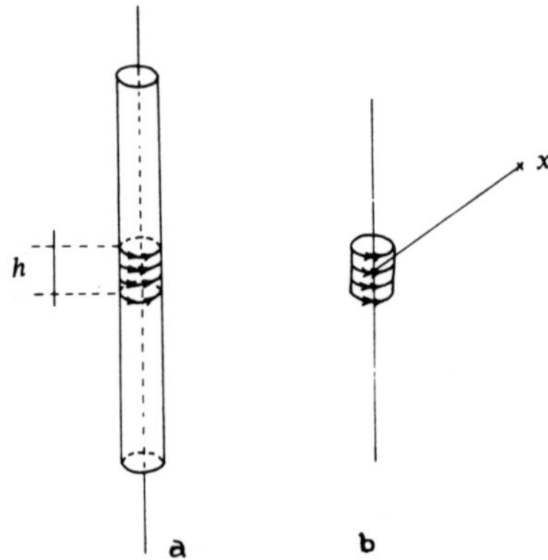

Fig. 34. (*a*) *Linear antenna.* (*b*) *Magnetic currents equivalent to the generator.*

For a thin wire, the effect of the extremities will be ignored and it will be assumed that at a point on the side z, $\vec{J}$ has the following form:

$$\vec{J}(x') = J(z')\vec{e_z}; \quad x' = \{a, \varphi, z'\} \tag{10}$$

where a is the radius of the antenna.

The current $I(z')$ is equal to $2\pi a J(z')$. On the axis, the total electric field is polarized along Oz. Since (1) is obeyed on the axis, it follows that:

$$\int_{S'} K(z, z')I(z')\, dz' = -E_{a,z}(z); \quad z \in [-l/2; +l/2] \tag{11}$$

with:

$$\left.\begin{aligned} K(z, z') &= \frac{2}{i\omega\varepsilon}\left[k^2 G(0, z; a, z') + \frac{\partial^2}{\partial z\, \partial z'} G(0, z; a, z')\right] \\ E_{a,z} &= -2\pi \int_{-(h/2)}^{+(h/2)} M_\varphi(z') \frac{a^2}{\rho} \frac{\partial}{\partial \rho} G(0, z; a, z')\, dz' \\ \rho &= [a^2 + (z - z')^2]^{1/2} \end{aligned}\right\} \tag{12}$$

where h is the height of the excitation zone.

Equation (11) constitutes an integral equation of the first type (Pocklington's equation), allowing the current to be calculated. For this equation, the point of observation is situated on the axis, with the current taken at a point on the surface of the antenna. Another, less strict, point of view, can be adopted in which only the metallic parts of the antenna S'_1 and S'_2 are considered. A current can then be found which creates on the surface of S a tangential component of the electric field, of the same type as that found with a slit:

$$\int_{S'_1 \cup S'_2} K(z, z') I(z')\, dz' = \begin{cases} -E_0;\ x \in S'' \\ \ \ 0;\ x \in S'_1 \cup S'_2 \end{cases} \tag{13}$$

or else the current that would be established on the two wires S'_1 and S'_2 when input current I_0 is imposed:

$$\left.\begin{aligned} &\int_{S'_1 \cup S'_2} K(z, z') I(z')\, dz' = 0; \quad x \in S'_1 \cup S'_2 \\ &I\left(\frac{h}{2}\right) = I\left(-\frac{h}{2}\right) = I_0 \end{aligned}\right\} \tag{14}$$

For the two previous integral equations, the observation point is situated on the surface of the antenna and the current is assumed to be located along the axis Oz. Thus the node remains regular.

In relation to the current in (7), the field can be calculated at any point in space. It is usual to ignore the contribution of the magnetic currents in the cut, if this is narrow, and if it is basically the far field that is being considered.

Pocklington's equation can be extended without difficulty to the case of linear antennas of any shape. The angular variations in current are assumed to be negligible in a straight section of the conductor, as are the possible transverse components, compared with the axial components (Figure 35):

$$\vec{J}(x') \simeq J(s')\, \vec{e}_{s'} \tag{15}$$

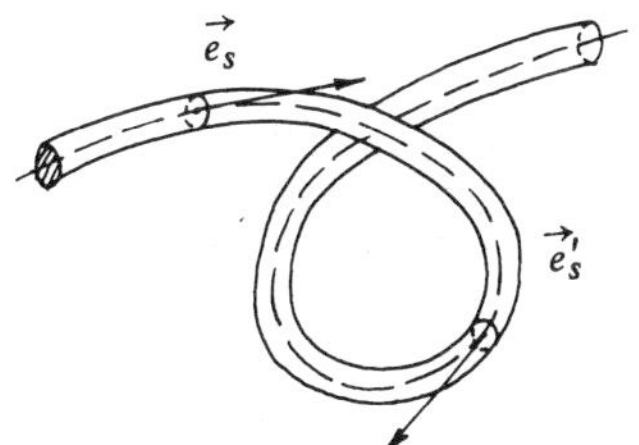

Fig. 35. *Wire antenna of irregular shape.*

A point on the antenna will be located by means of s, which corresponds to the centre-line of the antenna; $\vec{e}_s$ is a unit vector tangential to this path. It is easy to verify that the current $I(s')$ is the solution to the integral equation:

$$\int_0^S K(s, s')I(s')\, ds' = -E_{a,s}(s); \quad s \in [0, S] \tag{16}$$

with:

$$\left.\begin{aligned} K(s, s') &= \frac{2}{i\omega\varepsilon}\left[k^2\, \vec{e}_s \cdot \vec{e}_{s'} G(s, s') + \frac{\partial^2}{\partial s\, \partial s'} G(s, s')\right] \\ \partial_s A &= \vec{e}_s \cdot \text{grad } A \end{aligned}\right\} \tag{17}$$

The point of observation is taken on the centre-line and the current point is on the antenna surface.

2. Reception antennas All the previous equations referred to transmission mode. They can be extended to reception mode, since the transverse currents and variations in longitudinal currents in a straight section of the wire can be ignored. These approximations are acceptable for thin antennas.

Consider a linear antenna placed in a known field $\vec{E}_a(x)$. It is loaded at a point z_c with an impedance load Z, localized or distributed. If h is the width of the cut intended for the insertion of the load, the tangential component of the electric field satisfies the equation:

$$E_z(z_c) = -\frac{Z}{h} I(z_c) \tag{18}$$

Since the field is no longer zero along the whole extent of the antenna axis, the current can be determined from equation (11), with the node K replaced by the node K' such that:

$$K'(z, z') = K(z, z') + \frac{Z}{h}\left[U\left(z - z_c + \frac{h}{2}\right) - U\left(z - z_c - \frac{h}{2}\right)\right] \tag{19}$$

where U is Heaviside's function.

It is easy to extend this to the case of a wire antenna of any form, loaded with an arbitrary number of impedances.

3. Coupling The integral formulation also applies to coupling. It is assumed that an active linear antenna A_1, oriented along $\vec{e}_1$, radiates in the presence of another passive linear antenna A_2, oriented along $\vec{e}_2$ (Figure 36). To simplify the expression, A_2 is assumed to be short-circuited.

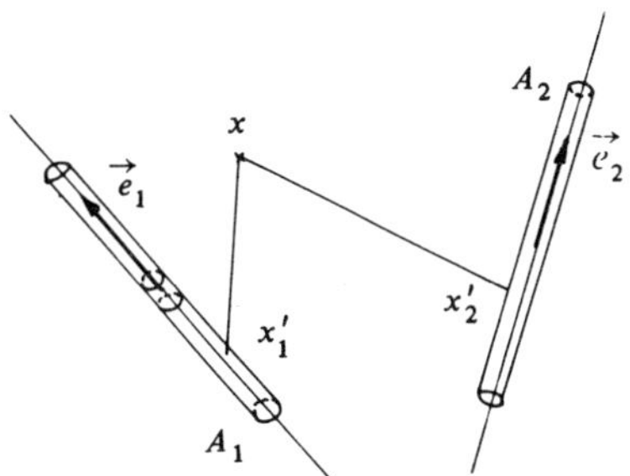

Fig. 36. *Coupling between two antennas.*

With the same notation as before, the field at any point is written as:

$$\vec{E}(x) = \vec{E}_a(x) - \frac{1}{i\omega\varepsilon}[k^2 + \text{grad div}]$$

$$\times \left\{ \int_{A_1} G(x, x')\vec{J}_1(x')\, dx' + \int_{A_2} G(x, x')\vec{J}_2(x')\, dx' \right\} \tag{20}$$

The currents I_1 and I_2 are deduced from the two coupled equations:

$$\begin{aligned} \vec{E}(x)\cdot\vec{e}_1 &= 0 \qquad x \in \text{axis of } A_1 \\ \vec{E}(x)\cdot\vec{e}_2 &= 0 \qquad x \in \text{axis of } A_2 \end{aligned} \tag{21}$$

These few examples show the generality of the integral formulation and its flexibility for dealing with very realistic situations.

4. Input impedance Calculation of the current does not lead only to that of the field radiated at any point it also allows the input impedance to be found:

$$Z_e = -\frac{\displaystyle\int_{z_e+(h/2)}^{z_e+(h/2)} E_z(x)\, dz}{I(z_e)} \tag{22}$$

where z_e is the excitation point on the antenna, and h the height of the cut.

This simple formula applies to cases where the modelling of the source is not too crude. In more general cases, the induced electromagnetic force method can be used, at the expense of longer calculations.

(b) Thick antennas with rotational symmetry

It is interesting to note that the Pocklington equation obtained during transmission is a specific case of a more general equation, applicable to thick antennas with rotational symmetry.

This antenna is completely described by its meridian $r = r_0(z)$, $z \in [z_a, z_b]$ (Figure 37). It is excited by a slot of width h made in the side $z = z_e$. The current

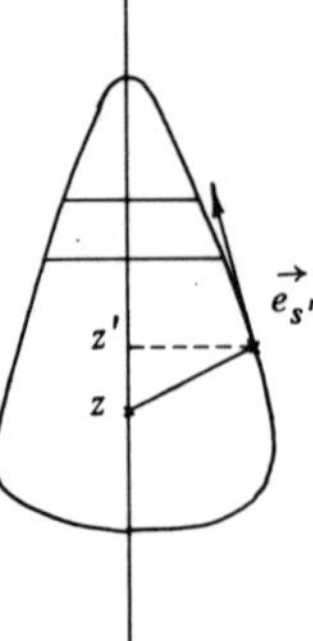

Fig. 37. *Thick antenna with rotational symmetry.*

density $\vec{J}$ is tangential to the surface, and contained in a meridian plane. The current $I(s)$ is defined by:

$$I(s) = 2\pi r_0(z) J \vec{e}_s \tag{23}$$

For any point inside S, the left-hand term of (7) is cancelled out. To determine $I(s)$, the procedure in principle is to write this condition for all the points inside S and for the three components of the total field.

In practice, this is unnecessary since each of the components deduced from (4) is an analytical function for any point inside S. It is enough, therefore, to express the cancellation of these components on an arc inside S, chosen to be as small as required. It is particularly useful to choose a segment of the antenna axis since, by symmetry, only the component in z remains.

The cancellation of this component gives rise to an equation comparable to (11) where the node is written as:

$$K(z, z') = \frac{2}{i\omega\varepsilon}\left[k^2 G(0, z; r_0(z'), z') + \frac{\partial^2 G}{\partial z\, \partial z'}(0, z; r_0(z'), z') + \frac{dr_0}{dz}\frac{\partial^2 G}{\partial r_0\, \partial z}(0, z; r_0(z'), z')\right] \tag{24}$$

This equation is due to Albert and Synge. As in the previous equations, the kernal K is regular. It is easy to prove that (11) can be derived in the case of a thin linear antenna, the effects of the extremities of which are ignored.

(c) Radiating apertures in perfectly-conducting planes

The excitation aperture can often be considered as being situated in an infinite perfectly conducting plane. Approximation is justified if the surface in which the aperture is made is of sufficient extent and if its curvature is not too great.

We have seen that the field in the half-space outside the plane can be expressed as a function of the tangential components of the single electric field in the aperture (Part 1, Chapter 1, section D).

$$\vec{E}(x) = 2 \operatorname{curl} \int_{\Pi''} G(x, x') \vec{M}(x')\, dx' \tag{25}$$

$$H(x) = -\frac{2}{i\omega\mu}[k^2 + \text{grad div}]\int_{\Pi''} G(x, x')\vec{M}(x')\,dx' \qquad (26)$$

The simplest approach consists of taking an approximate expression for $\vec{M}$ on Π'' as in Chapter 4. The field is deduced from this using a simple integration. It is also possible to determine the magnetic density that is established in Π'' for a given structure of excitation.

For example, let Π'' coincide with the straight section of a guiding structure (Figure 38) fed from a matched generator. In the half-space $z < 0$ the field admits a modal representation:

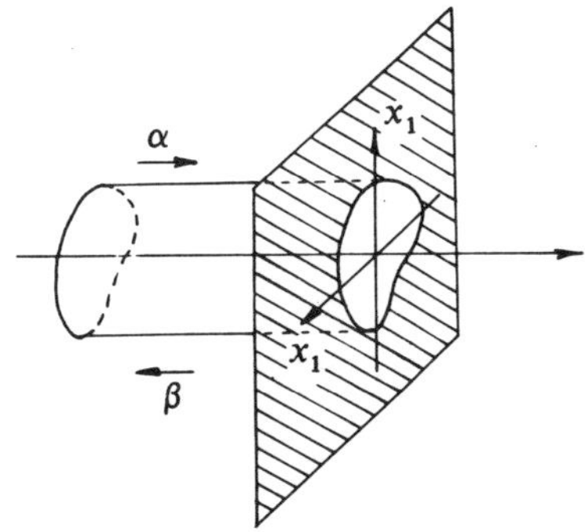

Fig. 38. *Waveguide structure opening into a conductive plane.*

$$\vec{E}(x) = \sum_n [\alpha_n e^{-\gamma_n z} + \beta_n e^{\gamma_n z}]\,\vec{e}_n(x_1, x_2) \qquad (27)$$

$$\vec{H}(x) = \sum_n [y_n][\alpha_n e^{-\gamma_n z} - \beta_n e^{\gamma_n z}]\,\vec{e}_z \wedge \vec{e}_n(x_1, x_2) \qquad (28)$$

where:
γ_n = propagation constant of the nth mode;
y_n = admittance of the nth mode;
$\vec{e}_n$ = specific vector associated with the electric field on the nth mode.

The quantities γ_n, y_n, $\vec{e}_n$ are assumed to be known for the structure considered. If the aperture is used for transmission, with the generator matched, then the coefficients α_m are known. The continuity of tangential components in the plane Π'' allows the two representations (25–26) and (27–28) to be linked:

$$\vec{J} = \vec{e}_z \wedge \vec{H}(z = 0^-) \qquad (29)$$

$$\vec{M} = \vec{e}_z \wedge \vec{E}(z = 0^-) \qquad (30)$$

By bringing in (24) a relationship can be deduced allowing the coefficients β_n to be determined as a function of α_n:

$$\sum_n \{\vec{F}(x_1, x_2) + y_n \vec{e}_n(x_1, x_2)\}\beta_n = \sum_n \{\vec{F}(x_1, x_2) - y_n \vec{e}_n(x_1, x_2)\}\alpha_n \qquad (31)$$

with:

$$\vec{F}(x_1, x_2) = \frac{2}{i\omega\mu} \vec{e}_z \wedge [k^2 + \text{grad div}] \int_{\Pi''} G(x_1, x_2, 0; x'_1, x'_2, 0)\, \vec{e}(x'_1, x'_2)\, dx'$$

The summation is extended to all incident modes for the left-hand term. For a monomodal structure, only a single term is taken. The coefficient β_n corresponding to this mode represents the reflection coefficient. It allows the reduced impedance to be calculated:

$$z_n = \frac{1 + \beta_n}{1 - \beta_n} \tag{32}$$

The above formulation can easily be extended to reception mode. In this case, (26) is completed as follows:

$$\vec{H}(x) = \vec{H}_a(x) + \bar{\vec{H}}_a(x) - \frac{2}{i\omega\mu_0} [k^2 + \text{grad div}] \int_{\Pi''} G(x, x')\vec{M}(x')\, dx' \tag{33}$$

$$\left.\begin{array}{ll} \text{with: } \vec{H}_a(x) = & \text{magnetic field created by the sources;} \\ \quad\;\; \bar{\vec{H}}_a(x) = & \text{magnetic field created by the image sources of the above} \\ & \text{relative to a plane } z = 0. \end{array}\right\} \tag{34}$$

If the receiver is matched, all the coefficients α_n are zero and the coefficients β_n are obtained using the system:

$$\sum_n \{\vec{F}(x_1, x_2) + y_n \vec{e}_n(x_1, x_2)\}\beta_n = \vec{H}_a(x_1, x_2, 0) + \bar{\vec{H}}_a(x_1, x_2, 0) \tag{35}$$

The presence of several guides opening onto the same plane can also be considered in the study of coupling phenomena between several apertures (as in arrays).

C: Introduction of numerical solution

1: Method of moments

All the equations considered can be reduced, more or less directly, to a functional equation of the type:

$$\mathscr{L}f = g \tag{36}$$

where f is an unknown function and g is a known one. Both are square integrable. $\mathscr{L}$ is a continuous linear operator. At a preliminary stage the function f is approximated by its projection $f^{(N)}$ on the first N elements of a complete base $\{f_n\}$:

$$f \simeq f^{(N)} = \sum_{n=1}^{N} \alpha_n f_n \tag{37}$$

The approximate equation deduced from this is projected onto the M first elements of a base $\{g_m\}$:

$$\langle \Sigma\, \alpha_n L(f_n) - g, g_m \rangle = 0; \quad m = 1, 2, \ldots, M \tag{38}$$

where $\langle\ ,\ \rangle$ denotes the usual scalar product.

The unknown coefficients α_n are the solution to the linear system:

$$[A][X] = [B] \tag{39}$$

where $[A]$ is a matrix $M \times N$ with general term $A_{mn} = \langle L(f_n), g_m \rangle$, and $[X]$ and $[B]$ are two column matrices with respective general terms $X_n = \alpha_n$ and $B_m = \langle g, g_m \rangle$. Since the coefficients A_{mn} and B_m are known, the coefficients α_n are deduced from them. If the extent of functions f and g is sufficiently regular, it can be shown, if the conditions are wide enough, that the solution $f^{(N)}$ converges as a norm towards f as N tends towards infinity.

It will be noted that for system (39) to admit a solution, it is necessary for $M \geqslant N$. In addition, the coefficients of the matrix $[A]$ and those of matrix $[B]$ depend only on the antenna (radiating or diffracting structure) and frequency for the former, and the field of excitation for the latter. There is, therefore, separation of the terms relating to the structure and those relating to the source.

The choice of bases f_n and g_m is arbitrary in principle, as long as they are complete. In practice, it determines the simplicity of the calculations of the coefficients A_{mn} and B_m, as well as the rapidity of convergence, and therefore the time of calculation and the computer memory capacity required. These functions can be defined on part of the support (echelons, triangles, etc) or on the whole of it (trigonometric functions, polynomials, etc).

There are many different types of numerical processing that can be applied to linear systems deduced by the method described above. Three of these are described below.

2: Direct numerical solution

System (39) can be solved numerically, once the calculation of the $M \times N$ coefficients of $[A]$ and the M coefficients of $[B]$ has been carried out. Usually, direct algorithms (Gauss–Jordan) are used, in preference to iterative algorithms, the convergence of which is not always certain. The system to be solved is usually of a rank varying from tens to hundreds. Generally, the rank increases with the dimension of the structure compared with the wavelength, and with the number of structures considered. When $M > N$, the overdetermined system can be solved using the least significant squares method. There is great interest in simultaneously calculating the various solutions $[X]_i$ corresponding to various excitations $[B]_i$, when the antenna structure and frequency remain unchanged. This results in a considerable saving in calculation time.

3: Determining the characteristic modes of the structure

As seen in the previous paragraph, the matrix $[A]$ is characteristic of the structure at a given frequency. Rather than characterize this structure by an array of $M \times N$ complex coefficients, it is sometimes preferable to do it using physical magnitudes. This is the aim of the method of characteristic modes, which was developed for cases where matrix $[A]$ results from the discretization of an equation resulting from an electric field. Let:

$$[A] = [U] + i[V] \tag{40}$$

where $[U]$ and $[V]$ are two real matrices. The characteristic current of the structure is that for which the matrix of unknown coefficients $[X_n]$ obeys:

$$[V][X_n] = \lambda[U][X_n] \tag{41}$$

One can show that the proper values λ_n and the proper vectors $[X_n]$ are real. The vectors constitute an orthogonal basis on the support. Finally, the definition of the characteristic currents using (41) implies the orthogonality of the radiation characteristics associated with them on a sphere of large radius centred on the antenna. These radiation characteristics can constitute useful bases in synthesis problems.

The solution $[X]$ corresponding to an excitation $[B]$ is simply expressed as a function of λ_n and $[X_n]$:

$$[X] = \sum_n \frac{1}{1 + i\lambda_n} \beta_n[X_n] \tag{42}$$

where:

$$\beta_n = \frac{[\tilde{X}_n][B]}{[\tilde{X}_n][U][X_n]} \tag{43}$$

$[\tilde{X}_n]$ designates the transposed matrix $[X]$. For structures the dimensions of which are not too large compared to the wavelength, the characteristic currents corresponding to the smallest specific values can be determined. When the excitation is modified, the numerator of the coefficients β_n can simply be recalculated.

In practice, the numerical determination of λ_n and $[X_n]$ is deduced from (41). This equation can be transformed into an equation with standard specific values by introduction of the matrix $[U]^{-1}[V]$.

4: Determining the poles in a complex plane

The extension to the complex plane of the previous harmonic analysis results in the matrix system:

$$[A(p)][X(p)] = [B(p)] \tag{44}$$

with $p = \sigma + i\omega$. This method was developed to characterize the structure with values that are here its poles in the complex plane. Fredholm's spectral equation theory shows that it is possible to find the solution $X(p)$, at least for a finite structure, in the form:

$$[X(p)] = \sum_n \frac{1}{p - p_n} [R(n)]\{[B(p)]\} + [F(p)] \tag{45}$$

where $[R(n)]$ is a matrix of residues independent of p and $[F(p)]$ is a matrix the elements of which are integer functions of p. The coefficients p_n are the zeros of the complex determinant of $[A(p)]$:

$$\det [A(p_n)] = 0 \tag{46}$$

On a practical level, the matrix $[R(n)]$ can be deduced from the proper vectors and convectors $[P_n]$ and $[Q_n]$, which are solutions to the equations:

$$\begin{aligned} [A(p_n)][P_n] &= 0 \\ [\tilde{A}(p_n)][Q_n] &= 0 \end{aligned} \tag{47}$$

It therefore appears that, in the development of the solution, the singularities arise partly from the structure—from the coefficients p_n—and partly from the excitation by $[B(p)]$. The poles p_n are distributed in the half-plane $\sigma > 0$, either on the real axis, or in conjugate pairs. The inverse Laplace transform makes it possible to go from (45) to the temporal response $[X(t)]$ of the structure with an arbitrary excitation $[B(t)]$. The term arising from $[F(p)]$ is eliminated generally by considerations of a physical type. Compared with other methods, this one requires a far greater numerical effort. In particular, finding zeros in the complex plane poses serious difficulties. Fortunately, the evaluation of $[X(t)]$ requires only a limited number of simple poles p_n to be determined. After having determined p_n, P_n and Q_n, the temporal response can be obtained very quickly, whatever the excitation.

D: An elementary example—the thin linear antenna

This example concerns a linear antenna of length l and radius $a \ll l$, in transmission or reception mode. Equation (11) allows the current $I(z)$ to be calculated. To this effect, the basic functions adopted are rectangular unit functions. The antenna is cut out in sections T_n $(n = 1, 2, \ldots, N)$ with average side z_n. Thus:

$$\Delta z = |z_{n+1} - z_n|$$

By collocation at M observation points $z_m(m = 1, 2, \ldots, M = N)$, a (39) type matrix system is obtained, and can be expressed in the form:

$$[Z][I] = [V]$$

with:

$$Z_{mn} = \frac{1}{\Delta z} \int_{T_n} K(z_m, z')\, dz'$$

$$I_m = I(z_n)$$

$$V_m = -E_a(z_m)\, \Delta z$$

The antenna appears as a multipole with N pairs of terminals. Each of these pairs corresponds to a section of antenna. The elements are identified with coupling impedances between the sections m and n.

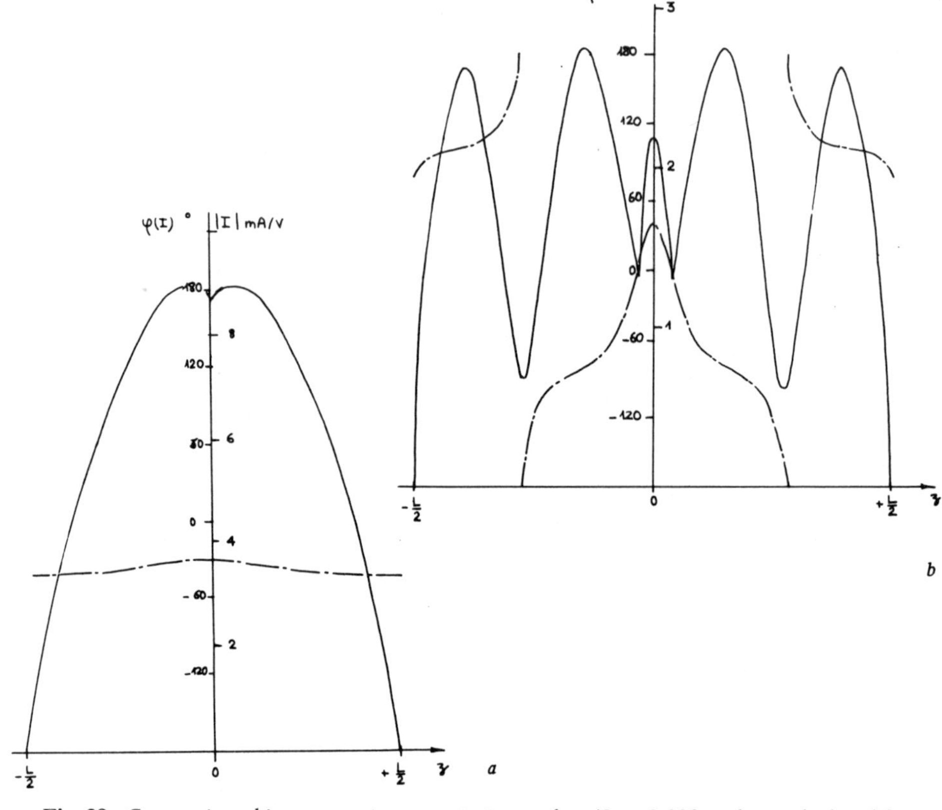

Fig. 39. *Current in a thin antenna in transmission mode. $a/L = 0.002$; values calculated for 24 sampling points. (a) $L = \lambda/2$; (b) $L = 2\lambda$.*

In transmission the generators can be simulated using layers—longitudinal or radial—of azimuthal magnetic currents. Taking into account the rapid decrease in the function $E^a(z)$, it will not cause any great error if the column matrix V is replaced with a matrix the elements of which are zero except for the one which corresponds to a section to which the generator is connnected. The antenna thus appears as a multipole in which the terminals are short-circuited, except for the one corresponding to the excitation section. The placing of the generator is arbitrary.

For reception, on the other hand, all the elements of the antenna are subject to comparable excitation fields. Also, all the terminals would be connected to a generator.

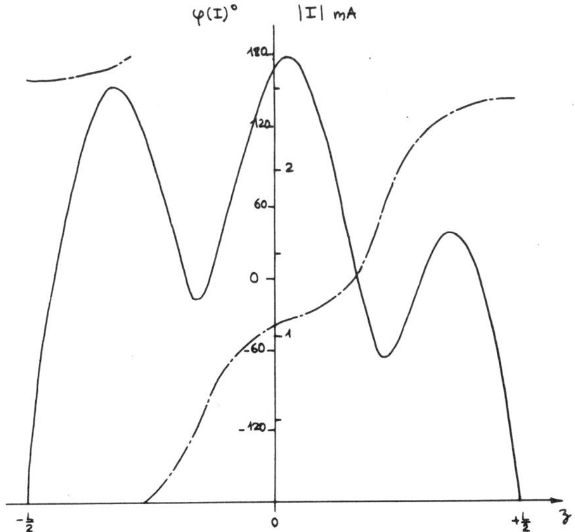

Fig. 40. *Current in a thin antenna illuminated by a plane wave; $a/L = 0.002$, $L = 1.5\lambda$, $N = 24$, $\theta = 45°$.*

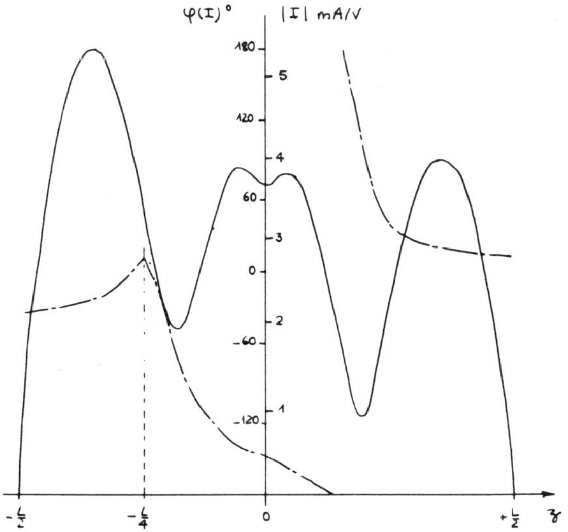

Fig. 41. *Current in a charged antenna operating in transmission mode. $a/L = 0.002$, $L = 1.5\lambda$, $N = 24$. The generator is situated at $z = L/4$, the charge is at $z = 0$.*

Whatever the mode of operation, it is easy to envisage the existence of a load impedance at any point on the antenna. The corresponding pair of terminals can simply be loaded with this impedance.

Note, finally, that the study of the antenna in the complex plane can be introduced quite naturally by analogy with circuit theory.

Figures 39 and 40 show some examples of current distributions.

Bibliography to Chapter 8

ALBERT G. E., SYNGE J. L. The general problem of antenna radiation and the fundamental integral equation *Quart. of Appl. Math.*, vol. **VI**, no. 1, p. 117 April 1948.

ANFREASEN M. G. Scattering from bodies of revolution, *IEEE Trans.*, AP-13, no. 2, p. 503, March 1965.

BOLOMEY J. Ch., TABBARA W. Numerical aspect of coupling between complementary boundary value problems, *IEEE Trans*, vol. AP-21, no. 3, May 1973, pp. 356–363.

BOLOMEY J. Ch. Méthode numériques en électromagnétisme, *Lecture Notes in computer Science*, vol. **11**, pp. 261–288, Springer 1974.

GALEJS J. *Antennas in Inhhomogeneous Media*, Pergamon Press, 1969.

HARRINGTON R. F. *Field Computation by Moment Methods*, Macmillan 1968.

MAUE A. W. Zur formulierung eines allgemeinen Bengungsproblems durch eine Integralgleichung, *Zeit. Physik*, Band **126**, 5601, 1949.

MIKHLIN S. G. *Integral Equations*, Pergamon Press.

MITTRA R. *Numerical and Asymptotic Techniques in Electromagnetics*, Springer 1975.

MITTRA R. *Computer Techniques for Electromagnetics*, Pergamon Press 1973.

MULLER C. *Grundprobleme der mathematischen Theorie elektromagnetischer Schwingungen*, Springer, 1957.

OSHIRO F. K., MITZNER K. M. Digital computer solution of three-dimensional scattering problems. *Second GISAT Symposium Proc.*, Bedford, Mass. Oct 1957.

Chapter 9

Experimental methods

A: Introduction

It is often not possible to take measurements on antennas in their real operating conditions, so the first experimental problem is to recreate these conditions in a laboratory. For non-directional antennas, the environment participates actively in the radiation. This is the case for a small antenna mounted on a vehicle, or for an antenna operating at low frequency close to the ground. It is therefore imperative to reproduce this environment as closely as possible in the laboratory. For highly directional antennas, on the other hand, the environment has little effect, and free space radiation conditions must then be simulated.

The measurements that may have to be made on a laboratory model concern the input impedance—or the reflection coefficient—as well as the radiated field. For the input impedance, the techniques used do not differ from those of normal circuits. For the measurement of electromagnetic fields, however, there are specific difficulties to be overcome. The technique used depends essentially on the quantities to be measured. The measurement of both near and far fields will be considered below.

B: Realization of a laboratory model

1: Principle of similarity

When the radiating system—here the antenna and its active environment—is very large or very small in relation to laboratory scale, it is very helpful to be able to create a reduced or enlarged scale laboratory model, allowing the conditions of real operation to be simulated. The model is said to be *quantitative* or *absolute* when the configuration of the field lines and the power level are identical. Very often *scale*, of *geometric*, models are used, in which only the configuration of the field lines are represented. Such models are adequate for the study of the main values in the radiating system, such as the reduced radiation characteristic, the directivity or the input impedance.

The possibility of realizing a scale model depends on the linearity of Maxwell's equations. Linearity of the radiation space is therefore indispensable, while its homogeneity is not. Consider p, the scale factor allowing any point $\vec{r}$ in the real system to have associated with it a point on the model $\vec{r}' = \vec{r}/p$. For a scale model, the configuration $\vec{E}'(\vec{r}')$, $\vec{H}'(\vec{r}')$ is linked to that of the real system

$\vec{E}(\vec{r})$, $\vec{H}(\vec{r})$ by the following relationships:

$$\vec{E}(\vec{r}) = \alpha \vec{E}'(\vec{r}')$$

$$\vec{H}(\vec{r}) = \beta \vec{H}'(\vec{r}')$$

where α and β are any two constants. Let γ be the ratio of the working frequency in the real system to that of the model:

$$\omega = \gamma \omega'$$

For a set of four constants p, α, β, γ, the problem consists of determining the electromagnetic constants $\varepsilon'(\vec{r}')$, $\mu'(\vec{r})$, $\sigma'(\vec{r})$ of the model as a function of those in the real system. The identification of Maxwell's equations expressed at the homologous points $\vec{r}$ and $\vec{r}'$, gives the following relationships:

$$\varepsilon'(\vec{r}) = \varepsilon(\vec{r}) \frac{\beta\gamma}{\alpha p}$$

$$\mu'(\vec{r}') = \mu(\vec{r}) \frac{\beta p}{\alpha\gamma}$$

$$\sigma'(\vec{r}') = \sigma(\vec{r}) \frac{\beta}{\alpha p}$$

It is easy to find this set of relationships by reasoning, as below, for the electric field:

$$\text{curl}_{M'} \vec{E}'(\vec{r}') = p \,\text{curl}_M \vec{E}'(\vec{r}') = \frac{p}{\alpha} \text{curl}_M \vec{E}(\vec{r})$$

In practice, there are constraints on the availability of usable materials that usually impose the choice of the three constants p, α, β. If the use of magnetic materials is ruled out, it follows that:

$$p\beta = \alpha\gamma$$

In addition, when propagation in the real system takes place partly in the air, one will decide also to use the air in the model. This precaution avoids the use of materials with large dimensions, which are inconvenient and generate losses. It follows that:

$$p\alpha = \beta\gamma$$

The two relationships above cannot be satisfied simultaneously unless:

$$\alpha = \beta$$

$$p = \gamma$$

The laws of correspondence between the characteristics of the real system and those of the scale model are thus obtained. The identity of the reduced characteristics and the gains can easily be proved. It may be helpful to note that for rcal non-dissipative materials the permeability and permittivity are not modified. For real dissipative materials, the conductivity is in inverse proportion to the lengths. This constraint limits the possibilities for making scale models. Among materials that can be used for this purpose are electrolytes, such as copper sulphate or carbon powders, which allow conductivity adjustable between a few m℧/m and a few k℧/m to be obtained.

2: Simulation of radiation conditions in free space

This problem is far simpler since it requires only the elimination of parasitic reflections from the walls, ground and ceiling of the laboratory. These reflections can be greatly attenuated by using absorbing materials with a very small reflection coefficient.

Two types of material are distinguished (Figure 42). The first are formed from a dielectric layer, the two faces of which have been suitably metal-plated so that the reflected field is cancelled out by interference. These structures are thin (with thickness only a fraction of the wavelength) and their band of operating frequencies is relatively reduced. Other materials can be assimilated into thick conductive layers, the conductivity profile of which is controlled to obtain progressive absorption, without reflection, of the incident wave. In practice, these are materials with constant conductivity and pyramidal profile. The height of the pyramids is about the same as the wavelength: the materials are thick but their band of operating frequencies is very wide and virtually unlimited for small wavelengths in the microwave range.

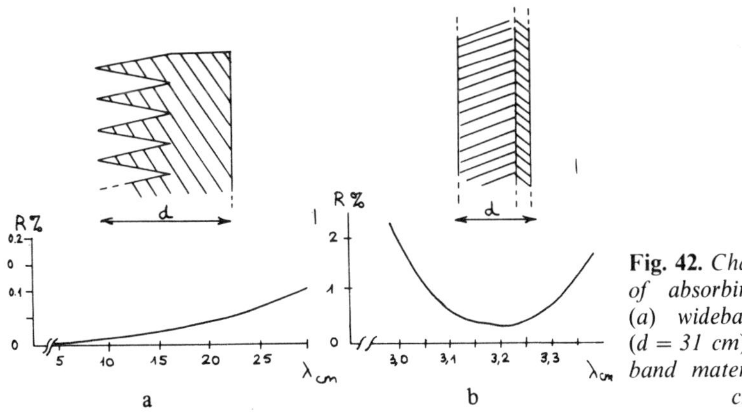

Fig. 42. *Characteristics of absorbing materials; (a) wideband materials (d = 31 cm); (b) narrow-band materials (d ≃ 3.5 cm).*

For both types of material, the reflection coefficient increases with the angle of incidence. The absorbing panels are directly applied to the laboratory walls (Figure 43). The rectangular structure of the anechoic chamber has gradually given way to pyramidal chambers, which are more economic and allow the most troublesome specular reflections to be avoided.

Fig. 43. *Anechoic chamber.*

C: Local measurement of electromagnetic fields

1: Measurement probes

The term probe is here applied to an antenna designed for the measurement of the electromagnetic field at a point in space. The electric field is measured using dipoles, the magnetic field using circular or square loops (Figures 44 and 45).

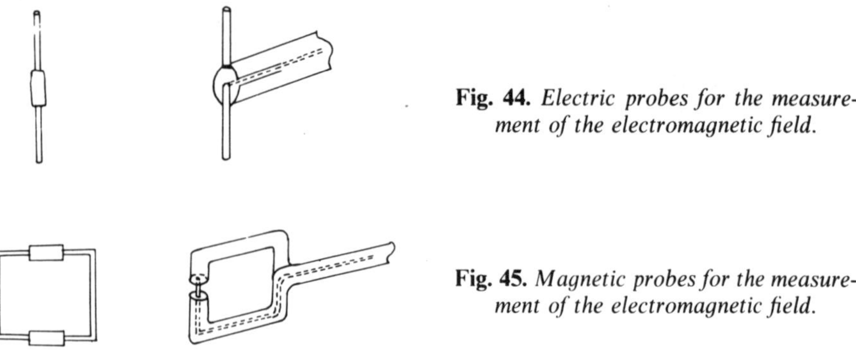

Fig. 44. *Electric probes for the measurement of the electromagnetic field.*

Fig. 45. *Magnetic probes for the measurement of the electromagnetic field.*

The available voltage at the terminals of the dipole is proportional to the average value of the electric field component parallel to it. For the loop, the problem is less simple. The available voltage is clearly proportional to the magnetic field component normal to the plane of the loop, but also to the variations in the electric field along the loop. This parasitic term is reduced more rapidly than the effective term when the surface area of the loop tends towards zero.

If the measurements are taken at points where the electric field undergoes variations comparable to those of the magnetic field, an open waveguide, sensitive to both types of field can be used.

The problem of local field measurement is a difficult one in that the introduction of the probe itself disturbs the initial distribution. This disturbance is inevitable and even necessary, since the measurement signal results from it. It is important to ensure, however, that it is very localized and that the presence of the probe does not modify the distribution of currents on the antenna that creates the field to be measured.

If the measurement is to be considered local, and the disturbance kept to a minimum, the probe must be as small as possible. The field to be measured should be seen as being broken down into an angular spectrum of plane waves. To ensure the correct representation of this field, the measurement probe must also be sensitive to all the plane waves—that is, it must be non-directional, and therefore small. There are two limiting factors to the reduction in probe size. The first is obvious: the smaller the probe, the less sensitive it is. The second is imposed by the need to have good decoupling relative to the components orthogonal to the measured component. This decoupling can only be obtained with a high length (electric probe)—or diameter (magnetic probe)—to thickness ratio. But for reasons of a practical nature, the thickness of the conductors that make up the probe cannot be reduced arbitrarily. To the intrinsic imperfections of probes can be added the more or less inevitable disturbance caused by the probe support. This must be designed to give the best compromise between disturbance and mechanical rigidity.

2: Measurement techniques

(a) Standard method

Independently of the problems mentioned above, it is important to take account of the disturbances introduced by the link between the measurement probe and the reception system. The most obvious method of measurement consists of linking the probe directly to a receiver that allows the signal to be amplified. This method, which can be referred to as standard, is very sensitive to this link, generally formed with a coaxial line (Figure 46).

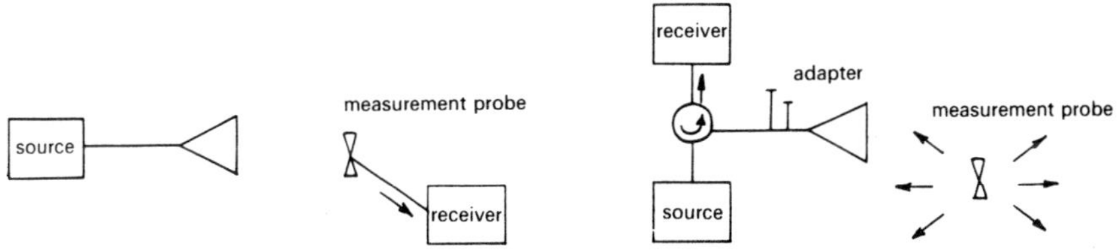

Fig. 46. *Field measurement by the standard method.*

Fig. 47. *Field measurement by the scattering method.*

(b) Diffusion method

The disturbance effect of a link can be eliminated by using the diffusion method, which consists of measuring the field re-radiated by the probe. Figure 47 shows a possible method for this. The antenna in question is transmitting. In the absence of the probe, the impedance transformer is adjusted so that no signal is detected by the receiver. The introduction of the probe destroys the equilibrium. The theory of reciprocity then allows the signal arriving at the receiver to be related to the field where the probe is situated. In this way, it is shown that the signal is proportional to the square of the field component to which the probe is sensitive.

In this form, there is no electrical link between the probe and the reception system, but the method is not very sensitive, because the re-radiated field is weak. To make the measurement signal easier to use, the re-radiated field is modulated. The modulation is simply obtained by loading the probe with an impedance which varies at the frequency of an electrical or optical signal. In the case of the former, the links necessary for the modulation are chosen to be highly resistive, to reduce their disturbance effect. This method requires substantially more complex equipment than the standard method. It is very useful for the exploration, in amplitude and phase, of fields with complex structures.

(c) Liquid crystal display

Liquid crystals are substances that change colour according to the temperature. All colours in the visible spectrum can be obtained with changes in temperature of only a few degrees. Liquid crystals can, therefore, be used to indicate locally the level of the electromagnetic field. To this end, they are deposited on a resistive film, the surface impedance of which is chosen equal to 377Ω. The local heating of the film, proportional to the intensity of the field, causes an increase in temperature and a consequent change in colour.

This gives a qualitative indication, rather than a measurement. This indication of the field distribution can be particularly useful when the time constant of the system is of the order of a second, as it is then possible to work in real time.

(d) Low-frequency antennas

As shown in Chapter 4, the distribution of the field radiated by small antennas has many analogies with quasistatic distribution. It is therefore possible to use the principle of rheographic tanks to explore the field lines.

D: Direct measurement

1: Stating the problem

This section is concerned with the description of measurement methods that allow the properties of a field far from an antenna to be characterized. These methods are direct in that the result is deduced more or less immediately from the measurement. Only two of the more common approaches will be discussed here. The first, and morc widcsprcad, uscs thc rcsponse of the antenna studied, in reception mode, when subjected to an incident plane wave. This gives the gain in a given direction or, more completely, the radiation characteristic in all directions. The second method, on the other hand, is based on the use of the antenna in transmission: the normal conditions of operation are deliberately disturbed so as to re-create nearby a field structure similar to that existing in the far field in normal conditions.

2: Antenna subjected to a plane wave

(a) Principle of measurement of the radiation characteristic

The principle of the method arises directly from the content of Chapter 7. The available voltage at the terminals of an antenna, operating in reception mode and subjected to a plane incident wave, can be written as:

$$U = A\vec{E}_0(\vec{u})\cdot\vec{F}_r(\vec{u}) \tag{1}$$

where A is a constant that depends only on the input impedance of the antenna and on that of the reception system connected at its terminals. $\vec{E}_0(\vec{u})$ is the field of the plane wave and $\vec{F}_r(\vec{u})$ is the reduced characteristic of the antenna in question.

For a given direction $\vec{u}$, the state of polarization of $\vec{F}_r$ can be determined by modifying that of the plane wave. It is helpful to use a rectilinear polarization, with variable inclination. The simplest method of revealing the angular variations in the characteristic is to rotate the antenna without modifying the direction of the plane wave. If the rotation does not take place exactly about the centre of the antenna in relation to which $\vec{F}_r$ is defined, the only result is an apparent shift without any practical importance, since it modifies both components of $\vec{F}_r$.

(b) Principle of gain measurement

In the same way, the power supplied by the antenna to the reception system, pointed in direction $\vec{u}$, is:

$$W'_r = \frac{\lambda^2}{4\pi}\frac{E_0'^2(u)}{\eta}\rho_c\rho_p(\vec{u})G(\vec{u}) \tag{2}$$

The value of the gain can be deduced from the measurement of power W'_r, once the polarization of the incident wave and the load have been adjusted, so that $\rho_c = \rho_p(\vec{u}) = 1$. It is often preferable to operate by comparison with an antenna, the structure of which is sufficiently simple for the gain to be calculated fairly precisely. Precautions must be taken if the antenna studied and the standard antenna do not have the same impedance or polarization.

(c) Practical solutions

The simplicity of this approach makes it attractive. The only difficulty is on a practical level: a locally plane wave must be used, over at least the whole extent of the antenna studied. In fact, the criteria for plane waves are rather vague. A maximal variation in amplitude of about one tenth of a decibel can be tolerated. For the phase, a limit of $\pi/8$ radian is fixed, more or less arbitrarily (Figure 48).

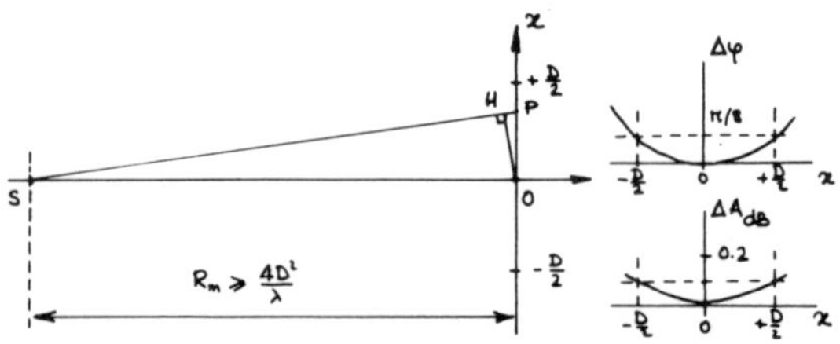

Fig. 48. *Distance criterion.*

To obtain this result, a preliminary solution consists of using the distant field of a source antenna. Once this antenna can be treated simply in terms of its phase centre, the distance of minimål separation R_m, by the phase criterion noted above, can be deduced. Since two rays from this centre, one incident on the centre of the antenna studied and the other incident or its side, present a geometric shift of less than $\lambda/16$, it is easy to obtain:

$$R_m \geqslant \frac{4D^2}{\lambda} \tag{3}$$

where D is the diameter of the aperture in the antenna studied. The amplitude constraint is generally less severe and it is important to ensure that the field does not have a longitudinal component. It is easy to see that relationship (3) can call for large separations; but the greater the distance between the two antennas, the more difficult it is to avoid parasitic reflections from the ground. For outdoor assemblies, these reflections can be considerably attenuated by using screens or by adopting, for the ground, a transverse configuration where ground reflections will be in a direction such that they cause no problems. For indoor assemblies, the use of absorbent materials provides a costly but technically satisfactory solution.

These difficulties have led to the development of more compact assemblies. An experimental assembly of this type is shown in Figure 49. The antenna considered is situated in the field close to a large reflector illuminated by an off-axis primary source. There are several practical problems.

Firstly the planarity of the reflected wave requires very strict tolerances on the state of the reflector surface. It can be shown that a deformation of about one hundredth of the wavelength over a non-negligible area can bring about errors of the order of a decibel in the gain measurement. This technological aspect explains the relatively recent development of the method. In addition, the application of (2) imposes a well-defined state of polarization over the whole surface of the antenna studied. But the polarization of the primary source is altered during reflection. The depolarization is weak about the axis and can possibly be reduced by increasing the focal length of the reflector. As shown in

Figure 49, the antenna in question can be subjected to parasitic radiation coming directly from the primary source or indirectly by reflection and diffraction from various supports, as well as from the sides of the reflector. The choice of an off-axis primary source and the rounding of the edges solve the problems concerning the reflector. For the rest, the use of absorbent materials is vital.

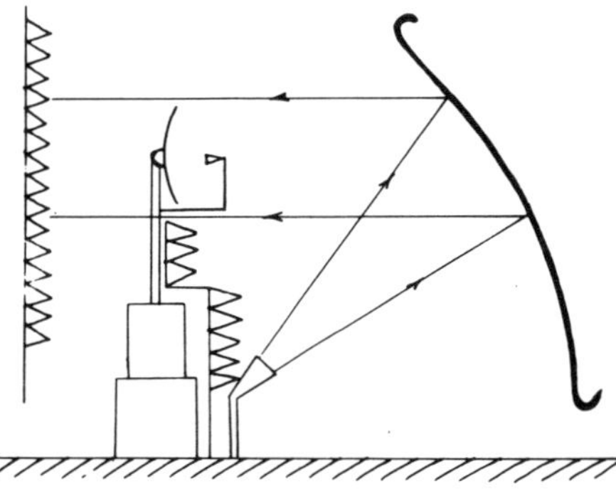

Fig. 49. *Compact assembly.*

Finally, the proximity of the antenna and the primary source may give rise to interactions absent from normal systems, where the antennas are distant from each other. In fact, this interaction has no great significance when the antenna under consideration intercepts only a small part of the reflected beam, of the order of one tenth. This is an important limitation with this assembly.

3: Disturbance of conditions of normal operation

This method is applied basically to focused systems and arrays. In normal operation, the various parameters (position of the primary source, phase shift between elements, etc) are adjusted so that the equiphasal surface of the radiated wave are as nearly plane as possible, so as to ensure focussing at infinity (Figure 50). By changing this, it is possible to obtain focus at a finite distance, far less than R_m.

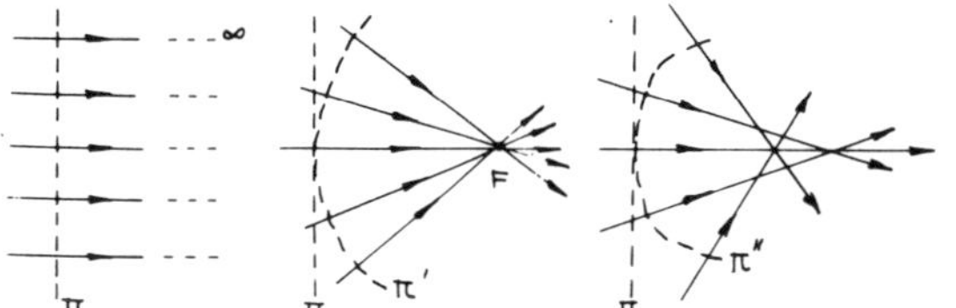

Fig. 50. *Unfocusing of focused systems.*

For arrays, the phase shift of each of the elements can be modified arbitrarily so as to give a focus without distortion (all the 'rays' issuing from the array converge at a single point on the axis). The shift can be carried out electronically or mechanically, by curving a linear array, for example.

In other cases, however, the modification of the phase law is accompanied by a distortion (aberrations) and it is necessary to take certain precautions. This is the case with a parabolic reflector, for example, in which the primary source is moved away from the reflector. (The 'rays' issuing from the primary source do not converge exactly at a single point on the axis, as would be the case for an elliptical reflector).

E: Indirect measurement

1: Principle

For transmission, the antenna can be replaced by equivalent sources distributed on a closed surface surrounding it. The sources are only the tangential components to this surface of the radiated field. These tangential components can be measured, and their inclusion in the general formulae allows the field to be calculated at any point outside the surface, and particularly at infinity. On a practical level, the methods recommended differ in the choice of source and the mode of representation adopted for the fields on this surface.

2: Practical solutions

A preliminary solution consists of choosing the external surface of the antenna as the measurement surface. For conductive antennas, the superficial densities of the current must be measured using electric or magnetic probes (Figure 51). This solution can be applied successfully to wire antennas, but less so to reflector antennas, because of the problems involved in moving the probe supports.

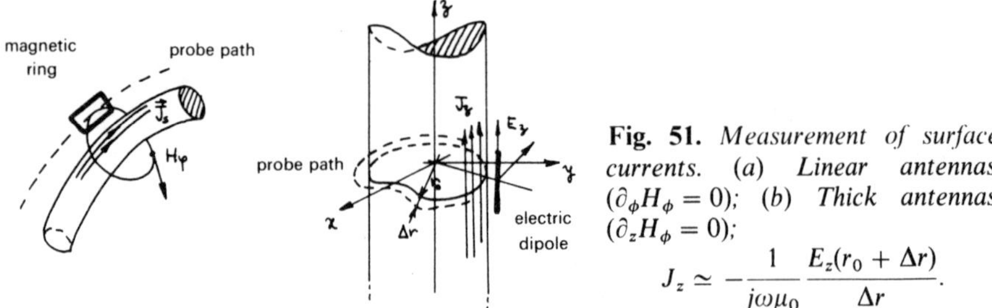

Fig. 51. *Measurement of surface currents.* (*a*) *Linear antennas* $(\partial_\phi H_\phi = 0)$; (*b*) *Thick antennas* $(\partial_z H_\phi = 0)$;

$$J_z \simeq -\frac{1}{j\omega\mu_0}\frac{E_z(r_0 + \Delta r)}{\Delta r}.$$

The natural choice, therefore, is that of a surface on which the movement of the probes will be easier. The surface must, in principle, be closed. In fact, any part of this surface on which the field can be considered to be zero is ignored. In practice the plane, sphere and cylinder will be found among the most extensively used surfaces. The plane surface is the simplest and can be adapted to use with large antennas. It also benefits from one of the properties mentioned previously: only the tangential components, either of the electric or magnetic field, have to be measured. If Π is the useful part of the plane—that in which the field takes significant values—then the radiated field can be calculated at any point in the half-space that does not contain the antenna, using the relationship:

$$\vec{E}(x, y, z) = 2 \operatorname{curl} \iint_{\Pi} G(x, y, z; x', y', 0)\vec{K}(x', y')\, dx'\, dy' \tag{4}$$

with:

$$\vec{K}(x, y) = \vec{e}_z \wedge \vec{E}(x, y, 0)$$

This relationship can also be rewritten by introducing the angular plane wave spectrum $\vec{F}(\alpha, \beta)$ creating the distribution $\vec{K}$ on Π:

$$\vec{F}(\alpha, \beta) = \frac{\lambda^2}{4\pi^2} \iint_{\Pi} \vec{E}_t(x, y, 0)\, e^{ik[\alpha x + \beta y]}\, dx\, dy \tag{5}$$

where $\vec{E}_r$ is the tangential component of $\vec{E}$.

By inverse Fourier transformation, the following is obtained:

$$\vec{E}_t(x, y, z) = \int_{-\infty}^{+\infty} \int_{-\infty}^{+\infty} \vec{F}(\alpha, \beta)\, e^{-i(k/\pi)[\alpha x + \beta y + \gamma z]}\, dx\, d\beta \tag{6}$$

$$E_z(x, y, z) = -\frac{1}{\gamma} [\alpha E_x(x, y, z) + \beta E_y(x, y, z)] \tag{7}$$

with $\gamma^2 = 1 - (\alpha^2 + \beta^2)$. The sampling theorem allows the measurement intervals Δx and Δy to be evaluated. If it is assumed that $F'(\alpha, \beta) = 0$ for $|\alpha| \geqslant \alpha_0$, $|\beta| \geqslant \beta_0$, then the electric field can be calculated from the measurements of the tangential components made in the plane $z = 0$ at the intervals:

$$\Delta x = \frac{\Pi}{k\alpha_0} \qquad \Delta y = \frac{\Pi}{k\beta_0} \tag{8}$$

If, in the plane $z = 0$, there are no more evanescent waves in z, then $\alpha_0^2 + \beta_0^2 \leqslant 1$. In this case, the maximal step is a half-wavelength. If, on the other hand, there are evanescent waves, $\alpha_0^2 + \beta_0^2 > 1$, the minimal attenuation Am can be fixed from the evanescent waves in the z plane:

$$Am = e^{-kz[\alpha_0^2 + \beta_0^2 - 1]1/2} \tag{9}$$

The maximal measurement interval is therefore:

$$\Delta S = \frac{\Pi}{k(\alpha_0^2 + \beta_0^2)^{1/2}} \tag{10}$$

The method extends to other simple surfaces, such as the cylinder or sphere. The electric field is then developed in cylindrical or spherical waves instead of being expressed as a spectrum of plane waves. On a practical level, the capacity and rapidity of the computer used limit the possibilities of this method.

Bibliography to Chapter 9

AUGUSTINE C. F. Field detector works in real time, *Electronics*, pp. 118–122, June 24, 1968.

BROWN J. A theoretical analysis of some errors in aerial measurements, *Proc. IEE*, February 1958, pp. 343–351.

HARRINGTON R. F. Small resonant scatterers and their use of field measurements, *IRE Trans*, vol. MTT–10, no. 3, 1962, p. 165.

HARRINGTON R. F. Field measurements using active scatterers, *IEEE Trans*, vol. MTT–11, no. 6, Oct. 1963, p. 454.

HU M. K. On measurements of microwave E and H field distribution by using modulated scattering methods, *IRE Trans*, vol. MTT–8, 1960, p. 295.

IIZUKA K. Photoconductive probe for measuring electromagnetic fields, *Proc. IEEE*, vol. 110, no. 2, 1963, p. 1746.

JENSEN F. Methods of determining far-field data from near-field measurements, *URSI Symposium Reprints*, Tbilissi, p. 541–550, Sept. 9–15, 1971.

JOHNSON R. C., ECKER H. A., HOLLIS J. S. Determination of far field antenna patterns from near field measurements, *Proc. IEEE*, vol. 61, no. 12. December 1973, pp. 1668–1694.

JOY E. B., PARIS D. T. Spatial sampling and filtering in near-field measurements, *IEEE Trans*, vol AP-20, no. 3, May 1972, pp. 253–261.

JUSTICE R., RUMSEY V. H. Measurement of electric field distribution, *IRE Trans.*, vol AP-3, no. 4, July 1955, pp. 177.

KERNS D. M. Correction of near-field antenna measurements made with an arbitrary but known measuring antenna, *Electronics Letters*, vol 6, no. 11, 28 May 1970, pp. 346–347.

KING R. W. P., WU T. T. *The Scattering and Diffraction of Waves*, Harvard University Press, 1959.

PLONSEY R. Surface current measurements with an electric probe, *IRE Trans.*, vol MTT-10, no. 3, May 1962, p. 214.

PUYNAUBERT J. Visualisation des ondes électromagnétiques hyperfréquence à l'aide de cristaux liquides, *l'Onde électrique*, vol. 52, no. 5, pp. 213–217, May 1972.

RICHMOND J. H. A modulated scattering technique for measurement of field distribution, *IRE Trans*, vol MTT-3 no. 4, July 1955, pp. 13.

SENIOR T. B. A. Surface roughness and tolerances in model scattering experiments, *IEEE Trans*, vol. AP-13, no. 4, July 1965, pp. 629.

SINCLAIR G. Theory of models of electromagnetic systems, *Proc. IRE*, vol 36, no. 11, November 1948, pp. 1364–1370.

VURAL A. M., CHENG D. K. A light-modulated scattering technique for diffraction field measurements, *Radio Science*, vol 68 D, no. 4, April 1964.

WHITESIDE H., KING R. W. P. The loop antenna as a probe, *IEEE Trans*, vol AP-12, no. 3, May 1964, pp. 291.

Appendices

Appendix 1: Bicomplex algebra

In a linear space of dimension n, and element $x = (x_1, \ldots, x_n)$ can be represented relative to the canonical base $e_1 = (1, 0, \ldots, 0)$, $e_2 = (0, 1, \ldots, 0), \ldots,$ by:

$$x = \sum_{i=1}^{n} x_i e_i \tag{1}$$

This space, when an internal multiplication $(x, y) \to xy$, associative and distributive relative to the linear combination, is defined on it, becomes an algebra of dimension n. The following can be written:

$$xy = \sum_{i,j=1}^{n} x_i y_j e_i e_j \tag{2}$$

relative to the products $e_i e_j$, themselves representable as elements by:

$$e_i e_j = \sum_{k=1}^{n} \alpha_{ijk} e_k \tag{3}$$

with the n^3 structure constants α_{ijk} linked so as to ensure associativity. The relationships (3) can be expressed in the form of a multiplication table.

The *bicomplex numbers* are elements of a commutative algebra of dimension 4. With a base $\{1, \text{i}, \text{j}, \text{k}\}$ the following representation is found:

$$x = a + bi + cj + dk \tag{4}$$

(a, b, c, d real). The multiplication table* is shown below:

	1	i	j	k
1	1	i	j	k
i	i	-1	k	$-j$
j	j	k	-1	$-i$
k	k	$-j$	$-i$	1

Note that $x = a + bi + j(c + di) = a + cj + i(b + dj)$.

* In part 1 any confusion between the element k and the propagation constant k is avoided by replacing the first by the product ij.

Bicomplex field

The formalism applies to sinusoidal fields. In the usual complex representation in $\mathbf{C}^3$:

$$\vec{E} = \vec{A}_E + i\vec{B}_E \tag{5}$$

$$\vec{H} = \vec{A}_H + i\vec{B}_H \tag{6}$$

The complex electric and magnetic fields can be associated in a bicomplex field:

$$\vec{C} = \vec{E} + j\vec{H} = \vec{A}_E + i\vec{B}_E + j\vec{A}_H + k\vec{B}_H \tag{7}$$

The transposition of operations for ordinary vector analysis can be achieved through the rectangular cartesian components:

$$C_l = E_l + jH_l, \quad l = 1, 2, 3$$

The bicomplex representation can also be used in the calculation of circular polarizations (with $E_x + jE_y$).

Appendix 2: Stationary phase method

This concerns the approximation of integrals in the form:

$$I = \int_a^b f(x)\, e^{i\tau g(x)}\, dx \tag{1}$$

with real τ and g, and complex f.

When τ (>0) is large, cos τg and sin τg have a large number of close amplitude oscillations in any interval where f varies slowly and where g is not stationary. It is assumed that the largest contribution comes from around the values $x = x_0$ such that $g'(x_0) = 0$. Consider the case of a single point x_0.

A simple calculation with limited expansions gives:

$$I \sim \sqrt{2\pi}\, e^{\pm i(\pi/4)} \frac{f(x_0)}{\sqrt{\pm g''(x_0)}} \frac{e^{i\tau g(x_0)}}{\sqrt{I}} \tag{2}$$

$\pm$ depending on whether $g''(x_0)$ is >0 or <0.

Fresnel's integral

The calculation uses Fresnel's integral,

$$F(x) = \int_0^x e^{i(\pi/2)t^2}\, dt = C(x) + iS(x) \tag{3}$$

encountered several times in this work. The following approximations are retained:

(*a*) x small: $$F(x) = x + i\frac{\pi}{6}x^3 + O(x^5) \tag{4}$$

(b) x large: $$F(x) = \frac{1+i}{2} - i\frac{e^{i(\pi/2)x^2}}{\pi x} + O\left(\frac{1}{x^3}\right), \quad x > 0 \tag{5}$$

with:

$$F(-x) = -F(x)$$

For $\mathbf{R}^2$

The technique is extended to integrals such that:

$$I = \int f(x, y)\, e^{i\tau g(x,y)}\, dx\, dy \tag{6}$$

In the specific case of a single point (x_0, y_0) where $g'_x = g'_y = g''_{xy} = 0$, the following is found:

$$I \sim 2\pi A \frac{f(x_0, y_0)}{\sqrt{|g''_{x_0^2}\, g''_{y_0^2}|}} \frac{e^{i\tau g(x_0, y_0)}}{\tau} \tag{7}$$

where $A = \pm i$ when $g''_{x_0^2}$ and $g''_{y_0^2}$ are both >0 or <0, and $A = 1$ when $g''_{x_0^2}$ and $g''_{y_0^2}$ have different signs.

Appendix 3: Distributions*

a: Definitions

(1) Distributions

A distribution is a continuous linear form T on a topological linear space ϕ of complex functions $\varphi(x)$ on $\mathbf{R}^n$; that is, with any φ, T associates a complex number $T(\varphi) = \langle T, \varphi \rangle = T \cdot \varphi$ such that

(a) $\forall a_i \in \mathbf{C}, \varphi_i \in \phi$: $$T\left(\sum_{i=1}^{k} a_i \varphi_i\right) = \sum_{i=1}^{k} a_i T(\varphi_i) \tag{1}$$

(b) $\lim_{N\to\infty} \varphi_N = \varphi$ implies: $$\lim_{N\to\infty} T(\varphi_N) = T(\varphi) \tag{2}$$

The set ϕ' of distributions on ϕ is, in its turn, structured in a linear space by:

$\forall \varphi \in \phi$: $$\left(\sum_{i=1}^{k} a_i T_i\right)(\varphi) = \sum_{i=1}^{k} a_i T_i(\varphi) \tag{3}$$

---

* This appendix is essentially a formulation. Here we restate the principal definitions and results of Part I (see Bibliography to Part I, [7], [30], [31], [36]).

and equipped with the convergence law $\lim_{N\to\infty} T_N = T$ if:

$$\forall \varphi \in \phi: \qquad \lim_{N\to\infty} T_N(\varphi) = T(\varphi) \tag{4}$$

(ϕ': topological dual of ϕ).

The support *supp* φ of a function is the smallest closed set in $\mathbf{R}^n$, outside which $\varphi = 0$.

A distribution T is zero in an open set Ω if $T(\varphi) = 0$, for any *supp* $\varphi \subset \Omega$. Its support *supp* T is the smallest closed set outside which $T = 0$.

(2) Ancillary spaces

The commonest are linear spaces of indefinitely differentiable functions, with 'well-behaved' properties at infinity and a convergence rule $\varphi_N \to \varphi$ (this can be reduced to $\varphi_N \to 0$).

Space $\mathscr{D}$ *supp* φ compact (that is, in the case of $\mathbf{R}^n$, closed and bounded). $\varphi_N \to 0$, if: (a) *supp* $\varphi_N \subset K$; (b) each series of partial derivatives $D^p\varphi_N \to 0$,† uniformly with regard to K.

Space $\mathscr{S}$ $\forall p \in N^n, k \in N$: $\lim_{|x|\to\infty} |x|^k D^p\varphi(x) = 0$
(with uniform convergence towards 0 of all sequences $|x|^k D^p\varphi_N(x)$.

Space $\mathscr{E}$ only indefinitely differentiable functions with convergence of all the series $D^p\varphi_N$, uniform on any compact set.

Implications: $\mathscr{D} \subset \mathscr{S} \subset \mathscr{E} \Leftrightarrow \mathscr{E}' \subset \mathscr{S}' \subset \mathscr{D}'$.

$\mathscr{D}'$ is the linear space of the usual distributions, $\mathscr{E}'$ is that of the compact support distributions, $\mathscr{S}'$ is that of the Fourier transformable distributions.

b: Measures

(1) Distribution measures

A (complex) additive numerical function of sets $\mu(\Delta)$ (such as volume, mass, electric charge, probability, etc), allows a Stieltjes integral to be associated with any $\varphi \in \mathscr{D}$:

$$T_\mu(\varphi) = \mu(\varphi) = \int \varphi(x)\mu(dx) = \int \varphi(x)\, d\mu \tag{5}$$

A distribution denoted T_μ, or simply μ if there is no ambiguity, is defined.

(2) Distributions with density $f(x)$

Let f be locally integrable—that is, on any compact set Δ: $\mu(\Delta) = \int_\Delta f\, dx$.

† $D^p = \partial^{|p|}/\partial x_1^{p_1} \ldots \partial x_n^{p_n}$, $p = (p_1, \ldots, p_n)$, $|p| = p_1 + \ldots p_n$.

With this is associated a distribution called T_f or **f** or even f such that:

$$T_f(\varphi) = \int \varphi(x) f(x)\, dx \tag{6}$$

This specific case, in which the usual functions are identified with distributions, is the key to the extension of standard analysis operations to distributions.

(3) Discrete distributions

With the complex charges $\lambda_1, \lambda_2, \ldots$ located respectively at the points $a_1, a_2, \ldots$, (5) can be reduced to:

$$\mu(\varphi) = \sum_i \lambda_i \varphi(a_i) \tag{7}$$

Dirac's delta measure, δ, is the specific case of the unit charge at the origin:

$$\delta(\varphi) = \varphi(0) \tag{8}$$

(7) is considered with respect to the measure $\sum_i \lambda_i \delta_{(a_i)}$, $\delta_{(a)}$ denoting the unit charge located at a.

(4) Simple layers

A charge distributed on a surface S (for example, in $\mathbf{R}^3$) with density $\sigma(x)$—any compact set Δ of $\mathbf{R}^3$ contains the charge $\int_{\Delta \cap S} \sigma\, dS$ on S—defines the measure $\sigma\delta_S$, such that:

$$\langle a\delta_S, \varphi \rangle = \int_S \sigma(x)\varphi(x)\, dS \tag{9}$$

δ_S represents the uniform layer $\sigma = 1$.

(5) Threads

A curve C carrying a linear density $\lambda(x)$ defines $\lambda\delta_C$ such that

$$\langle \lambda\delta_C, \varphi \rangle = \int_C \lambda(x)\varphi(x)\, ds \tag{10}$$

c: Other examples of distributions

(1) Multipoles

The measure $\mu_h = (\delta_{(b)} - \delta_{(a)})/h$, $b = a + \vec{\alpha}h$ ($\vec{\alpha}$ is a unit vector in the direction ab) converges (in $\mathscr{D}'$) for $h \to 0$ towards the distribution T such that:

$$T(\varphi) = \lim_{h \to 0} \frac{\varphi(b) - \varphi(a)}{h} = \partial_\alpha \varphi(a) \tag{11}$$

T is the unit dipole, with axis α, at a.

With N unit vectors $\alpha_1, \ldots, \alpha_N$ issuing from a, whether distinct or not, a $2N$-pole can be defined:

$$T(\varphi) = \partial^N_{\alpha_1 \ldots \alpha_N} \varphi(a) \tag{12}$$

(2) Double layer
As in (g), a surface density $\tau(x)$ on S defines a distribution, $\partial_n(\tau\delta_S)$, by:

$$\langle \partial_n(\tau\delta_S), \varphi \rangle = -\int \tau(x) \partial_n \varphi(x)\, dS \tag{13}$$

(3) Finite parts
The integral (6) is divergent if $f(x)$ ceases to be integrable around a point—for example, the origin. The idea of the finite part leads to the definition of a new distribution, denoted Pff, by:

$$\langle Pff(x), \varphi \rangle = PF \int f(x)\varphi(x)\, dx = \lim_{\varepsilon \to 0} \left[\int f\varphi\, dx - g(\varepsilon) \right] \tag{14}$$

when there is an infinite function $g(\varepsilon)$, like the integral, with $1/\varepsilon$ and such that the last member exists.

Examples:
(a) **R**. If $u(x)$ is Heaviside's step function:

$$\left\langle Pf\frac{u(x)}{x}, \varphi \right\rangle = PF \int_0^\infty \frac{\varphi(x)}{x}\, dx = \lim_{\varepsilon \to 0} \left[\int_\varepsilon^\infty \frac{\varphi(x)}{x}\, dx + \varphi(o) \log \varepsilon \right] \tag{15}$$

and:

$$Pf\frac{u(-x)}{x} \quad \text{by} \quad \left(\int_{-\infty}^{-\varepsilon} \right) - \varphi(o) \log \varepsilon$$

Thus, in particular, by addition and subtraction:

$$\left\langle Pf\frac{1}{x}, \varphi \right\rangle = \lim_{\varepsilon \to 0} \left[\int_{-\infty}^{-\varepsilon} + \int_\varepsilon^\infty \frac{\varphi(x)}{x}\, dx \right] = PV \int_{-\infty}^{\infty} \frac{\varphi(x)}{x}\, dx \tag{16}$$

(PV: Cauchy's principal value.)

$$\left\langle Pf\frac{1}{|x|}, \varphi \right\rangle = \lim_{\varepsilon \to \infty} \left[\int_\varepsilon^\infty \frac{\varphi(x) + \varphi(-x)}{x}\, dx + 2\varphi(o) \log \varepsilon \right] \tag{17}$$

(b) $\mathbf{R}^3$. $Pf\frac{1}{r^\alpha}$, $r = |x|$.

Thus:

$$\left\langle Pf\frac{1}{r^3}, \varphi \right\rangle = \lim_{\varepsilon \to 0} \left[\int_{r > \varepsilon} \frac{\varphi(x)}{r^3}\, dx - 4\pi\varphi(o) \log \varepsilon \right] \tag{18}$$

$$\left\langle Pf\frac{1}{r^4}, \varphi \right\rangle = \lim_{\varepsilon \to 0}\left[\int_{r>\varepsilon} \frac{\varphi(x)}{r^4}\,dx - 4\pi\frac{\varphi(o)}{\varepsilon}\right] \tag{19}$$

Pf is unusable for $\alpha < 3$.

d: Vector distributions

The most rapid solution is to use the intermediary of rectangular cartesian coordinates in $\mathbf{R}^3$. A vector distribution $\vec{T}$ is a triplet (T_1, T_2, T_3). $T_i \in \mathscr{D}$. Thus:

$$\vec{T}(\varphi) = \langle \vec{T}, \varphi \rangle = (T_1 \cdot \varphi, T_2 \cdot \varphi, T_3 \cdot \varphi) \tag{20}$$

Variant: With $\vec{\varphi} = (\varphi_1, \varphi_2, \varphi_3) \in \mathscr{D}^3$, it is also possible to form:

$$\vec{T}(\vec{\varphi}) = T_1 \cdot \varphi_1 + T_2 \cdot \varphi_2 + T_3 \cdot \varphi_3 \tag{21}$$

Vector layers

A usual field $A = (A_1, A_2, A_3)$ attached to a surface S defines the distribution:

$$\vec{A}\delta_S = (A_1\delta_S, A_2\delta_S, A_3\delta_S)$$

(triplet of simple 'component' layers) with:

$$\langle \vec{A}\delta_S, \varphi \rangle = \int_S \vec{A}(x)\varphi(x)\,dS \tag{22}$$

Example:

$\vec{A} = \vec{n}$ (the unit normal on S):

$$\langle \vec{n}\delta_S, \varphi \rangle = \int_S \varphi\vec{n}\,dS \tag{23}$$

Double vector layers

Distribution $\partial_n(\vec{A}\delta_S)$:

$$\langle \partial_n(\vec{A}\delta_S), \varphi \rangle = -\int_S \vec{A}(x)\partial_n\varphi(x)\,dS \tag{24}$$

Example:

$$\langle \partial_n(\vec{n}\delta_S), \varphi \rangle = -\int_S \partial_n\varphi\,dS\vec{n} \tag{25}$$

e: Operations on the distributions

(1) Procedure

Example: differentiation

Consider $f(x)$, locally integrable, with locally integrable partial derivatives $\partial_i f$. Integrating by parts:

$$\int \partial_i f\varphi\,dx = -\int f\,\partial_i\varphi\,dx$$

which suggests the definition of $\partial_i T$ by:

$$\forall \varphi \in \mathscr{D}: \langle \partial_i T, \varphi \rangle = -\langle T, \partial_i \varphi \rangle \tag{26}$$

and furthermore:

$$\langle D^p T, \varphi \rangle = (-1)^{|p|} \langle T, D^p \varphi \rangle \tag{27}$$

Thus any distribution is indefinitely differentiable in the above sense, which coincides, for $T = T_f$, with that of standard analysis.

A transposed operator ${}^tA: \mathscr{D} \to \mathscr{D}$ is tried to correspond to an operator $A: \mathscr{D}' \to \mathscr{D}'$, with the former defined by:

$$\langle AT, \varphi \rangle = \langle T, {}^tA\varphi \rangle \tag{28}$$

such that for $T = T_f$:

$$\int Af\varphi \, dx = \int f\, {}^tA\varphi \, dx \tag{29}$$

The operator A is continuous when $T_N \to T$ (in $\mathscr{D}'$) entails $AT_N \to AT$ (in $\mathscr{D}'$).

2: Examples:

(a) Translation (or shifting) $\tau_a T$ corresponds to $\tau_a f(x) = f(x - a)$

$$\langle \tau_a T, \varphi \rangle = \langle T, \tau_{-a} \varphi \rangle \tag{30}$$

(b) Symmetry If $\check{f}(x) = f(-x)$, the following is defined:

$$\langle \check{T}, \varphi \rangle = \langle T, \check{\varphi} \rangle \tag{31}$$

T is symmetrical (resp. antisymmetrical) if $T = T$ (respectively $-T$).

(c) Differentiation See (26) and (27). The Laplacian ΔT is retained:

$$\langle \Delta T, \varphi \rangle = \langle T, \Delta \varphi \rangle \tag{32}$$

(d) Multiplication by a function Consider $\alpha \in \mathscr{E}$:

$$\langle \alpha T, \varphi \rangle = \langle T, \alpha \varphi \rangle \tag{33}$$

Notably:

$$\alpha \delta = \alpha(o) \delta \tag{34}$$

(e) Fourier transformation $\varphi \in \mathscr{S}$, $\mathrm{T} \in \mathscr{S}'$. Parseval's equality is transposed:

$$\langle \hat{T}, \varphi \rangle = \langle T, \hat{\varphi} \rangle \tag{35}$$

(f) Applications of differentiation

1: *Discontinuous functions.*—(a) *On* **R**. Let $f(x)$ be differentiable as many times as necessary in the open intervals $]a_i, a_{i+1}[$. The possible finite discontinuity of a derivative at $x = a_i$ is:

$$f^{(p)}(a_i^+) - f^{(p)}(a_i^-) = [f^{(p)}]_{a_i} \tag{36}$$

then:

$$T'_f = T_{f'} + \sum_i [f]_{a_i} \delta_{(a_i)}$$

- -

$$T_f^{(p)} = T_{f^{(p)}} + \sum_i \left(\sum_{k=0}^{p-1} [f^{(k)}]_{a_i} \right) \delta_{(a_i)}^{(p-k-1)} \tag{37}$$

which generalizes $T_u^{(p)} = \delta^{(p-1)}$ (where u is Heaviside's step function).

(*b*) In $\mathbf{R}^3$. Let S be a closed surface, piecewise regular, separating the internal open set Ω from an external open set Ω'. The unit normal $\vec{n} = (n_1, n_2, n_3)$ is oriented towards the latter. If, however, $f(x)$, continuously twice differentiable in each of the open sets, is discontinuous across S, the discontinuity is defined at any point x of S as:

$$[f]_S = f(x^+) - f(x^-) \tag{38}$$

The same calculation as for R gives:

$$\partial_i T_f = \partial_i f = T_{\partial_i f} + [f]_S n_i \delta_S \tag{39}$$

(the 'usual' derivative in S^C plus the simple layer of density $[f]_{S^{n_i}}$)

$$\partial^2_{ij} f = T_{\partial^2_{ij} f} + [\partial_i f] n_i \delta_S + \partial_i([f]_S n_i \delta_S) \tag{40}$$

In particular, by writing Δf T_{Δ_f}:

$$\Delta f = \Delta f + [\partial_n f]_S \delta_S + \partial_n([f]_S \delta_S) \tag{41}$$

(usual laplacian + simple layer $[\partial_n f]_S$ + double layer $[f]_S$)

A specific case: $f = 0$ in Ω'. Note that $f(x^-) = f_S$:

$$\partial_i f = \partial_i f - f_S n_i \delta_S \tag{42}$$

$$(\partial_i f \text{ meaning } T_{\partial_i f})$$

$$\Delta f = \Delta f - (\partial_n f_S)\delta_S - \partial_n(f_S \delta_S) \tag{43}$$

the latter is linked to Green's formula.

Application: The following results are deduced from (43):

$$\Delta \frac{1}{r} = -4\pi\delta \tag{44}$$

$$\Delta \frac{e^{-ikr}}{r} = -k^2 \frac{e^{-ikr}}{r} - 4\pi\delta \tag{45}$$

which express the fact that $-\frac{1}{4}\pi r$ and $-e^{-ikr}/4\pi r$ are elementary solutions to the equations $\Delta U = f$ and $(\Delta + k^2)U = f$ respectively.

Note:
These formulae are valid only in $\mathbf{R}^3$. In $\mathbf{R}^2$ the corresponding elementary solutions would be $\Delta \log r/2\pi$ and $iH_0^{(2)}(kr)/4$ (where $H_0^{(2)}$ is Bessel's function of the third type, or Hankel's function of order 0).

2: *Multipoles.*—These can be defined directly by differentiation of $\delta_{(a)}$. Then $M\partial_\alpha\delta_{(a)}$ is the dipole at a, of moment M, and axis α, while $M_N\, \partial^{|N|}_{\alpha_1 \ldots \alpha_N}\, \delta_{(a)}$, where M_N is a tensor moment, will represent a $2N$-pole.

3: *Vector differentiations.*—(*a*) *Extension of usual operators in* $\mathbf{R}^3$. For T, α scalar distribution:

$$\text{grad } T = (\partial_1 T, \partial_2 T, \partial_3 T) \tag{46}$$

and for $\vec{T}$, a vector distribution:

$$\text{div } \vec{T} = \partial_1 T_1 + \partial_2 T_2 + \partial_3 T_3 \tag{47}$$

$$\text{curl } \vec{T} = (\partial_2 T_3 - T_2, \partial_3 T_1 - \partial_1 T_3, \partial_1 T_2 - \partial_2 T_1) \tag{48}$$

which verify the familiar properties. The extension to bicomplex fields is evident.

(*b*) *Discontinuities of fields.*—From (39), the following are found:

$$\begin{aligned} \text{grad } \mathbf{f} &= \text{grad} f + \vec{n}[f]_S\, \delta_S \\ \text{div } \vec{\mathbf{A}} &= \text{div } \vec{A} + n \cdot [\vec{A}]_S\, \delta_S \\ \text{curl } \vec{\mathbf{A}} &= \text{curl } \vec{A} + \vec{n} \wedge [\vec{A}]_S\, \delta_S \end{aligned} \tag{49}$$

Application: Field 'sweeped-out' outside S. The indicator $1_\Omega(x)$ is used to denote the internal domain. The following is obtained, for example:

$$\Delta(1_\Omega \vec{A}) = 1_\Omega\, \Delta\vec{A} - \partial_n \vec{A}\, \delta_S - \partial_n(\vec{A}\, \delta_S) \tag{50}$$

with the derivative of the vector field on S in the direction of the normal vector:

$$\partial_n \vec{A} = (\vec{n} \cdot \text{grad})\vec{A} \tag{51}$$

This is the vector with components $\vec{n} \cdot \text{grad } A_i = \partial_n A_i$.

(*c*) *Differentiation of layers.*—Scalar layer $f\delta_S$:

$$\langle \partial_i(f\, \partial_S), \varphi \rangle = -\langle f\delta_S, \partial_i\varphi \rangle = -\int_S f\, \partial_i\varphi\, dx \tag{52}$$

Hence, in particular:

$$\langle \text{grad } (f\, \delta_S), \varphi \rangle = -\int_S f \text{grad } \varphi\, dS \tag{53}$$

Then:

$$\langle \Delta(f\delta_S), \varphi \rangle = \int_S f\,\Delta\varphi\,dS \tag{54}$$

- Vector layer $\vec{A}\delta_S$. From the derivatives of the components:

$$\begin{aligned} \langle \operatorname{div}(\vec{A}\,\delta_S), \varphi \rangle &= -\int_S \vec{A}\cdot \operatorname{grad}\varphi\,dS \\ \langle \operatorname{curl}(\vec{A}\,\delta_S), \varphi \rangle &= -\int_S \vec{A}\wedge \operatorname{grad}\varphi\,dS \end{aligned} \tag{55}$$

(g) Convolution

1: *Usual functions.*—Subject to the existence of the integral:

$$f * g(x) = \int f(x')g(x - x')\,dx' = \int f(x - x')g(x')\,dx' = g * f(x) \tag{56}$$

Example—$f(x)$ density of bounded support in $\mathbf{R}^3$: $f * (1/|x|)$ is a newtonian potential, $f*(e^{-ik|x|}/|x|)$ is a Lorentz retarded potential.

It is easy to prove that:

$$\langle f * g, \varphi \rangle = \iint f(x')g(x - x')\varphi(x)\,dx'\,dx = \langle f, \check{g} * \varphi \rangle \tag{57}$$

2: *Regularized distribution.*—$\forall \alpha \in \mathscr{D}$, $T * \alpha$ is defined by:

$$\langle T * \alpha, \varphi \rangle = \langle T, \check{\alpha} * \varphi \rangle \tag{58}$$

Explicitly:

$$T * \alpha(x) = \langle T_{x'}, \alpha(x - x') \rangle \tag{59}$$

$T * \alpha$, regularized distribution T by the function $\alpha(x)$ is an indefinity differentiable function.

3: *Distributions.*—A distribution $T * U$ is defined by:

$$\langle T * U, \varphi \rangle = \langle T, \check{U} * \varphi \rangle \tag{60}$$

Cases of existence:
1: One at least of the supports compact.
2: The two supports in a translated convex cone Γ (such that trihedral angle $x_i \geqslant 0$, cone of revolution: $x_3 \geqslant \sqrt{x_1^2 + x_2^2, \ldots}$). T, $U \in \mathscr{D}'_\Gamma$ is written. The case, on $\mathbf{R}$, of the distributions with support contained in $x \geqslant 0(\mathscr{D}'_+)$ is important in signal theory.

The product is commutative, when it exists. It is associative in the following cases:
1: All the supports, except one at most, are compact.
2: In $\mathscr{D}'_\Gamma$.

4: *Properties.*—The formula proved in definition (50),

$$D^p\delta_{(a)} * T = D^p(\tau_a T) \tag{61}$$

includes the particular fundamental cases:

$$\begin{aligned} \delta * T &= T * \delta \\ \delta_{(a)} * T &= T * \delta_{(a)} = \tau_a T \\ \partial_i \delta * T &= T * \partial_i \delta = \partial_i T \\ D^p\delta * T &= T * D^p\delta = D^r T \end{aligned} \tag{62}$$

which signify that δ is a convolution operation and that translation and differentiation are specific convolutions.

Notably: $\Delta T = \Delta\delta * T$.

In addition, there are properties of permutation:

$$\begin{aligned} \tau_a(T * U) &= \tau_a T * U = T * \tau_a U \\ D^p(T * U) &= D^p T * U = T * D^p U \\ D^p(T * U) &= D^k T * D^{p-k} U \end{aligned} \tag{63}$$

5: *Continuity.*—$T_k \to T$ entails: $T_k * U \to T * U$.

(a) Convergence of functions to any distribution. There are series $\{\psi_k\}$ in $\mathscr{D}$ (for example, $\theta(kx)/\int \theta(kx)\,dx$, $\theta(x) = \exp(1/|x|^2 - 1)$ which converge towards δ in $\mathscr{D}'$. The regularized $T * \psi_k$ converge towards T.

In this way the newtonian or retarded potentials of the single and double layers are found:

$$f\delta_S * (1/r), f\delta_S * (e^{-ikr}/r), \partial_n(g\delta_S) * 1/r, \partial_n(g\delta_S) * (e^{-ikr}/r).$$

(b) Filter theorem. U fixed in $\mathscr{D}'$ (resp. $\mathscr{D}'_\Gamma$) defines a continuous linear mapping $\mathscr{E}' \to \mathscr{D}'$ (resp. $\mathscr{D}'_\Gamma \to \mathscr{D}'_\Gamma$): $T \to T * U$, permutable with the translation. The converse is true: any mapping $\mathscr{L}$: $\mathscr{E}' \to \mathscr{D}'$ having those characteristics is the convolution:

$$\mathscr{L}(T) = \mathscr{L}(\delta) * T \tag{64}$$

The case of $\mathbf{R}$ constitutes a basic theorem: with distributions, all linear, continuous, stationary transmission systems (filters) are convolution systems.

The case of $\mathbf{R}^2$ is encountered in optical imaging and in signal-processing antennas (Volume 2, Part 1).

(c) Convergence of measures. Conversely, any continuous function with compact support is the limit of a series of linear combinations of Dirac delta measures (this is in effect the calculation of Riemann's integral), so from (a) is deduced the convergence of such combinations towards a distribution T.

6: *Vector convolutions.*—With support conditions adapted from the above, $\vec{T} * U = (T_1 * U, T_2 * U, T_3 * U)$ is defined; then, with two vector distributions, the scalar distribution

$$\vec{T} \overset{*}{\cdot} \vec{U} = T_1 * U_1 + T_2 * U_2 + T_3 * U_3 \tag{65}$$

and the vector distribution

$$\vec{T} \underset{\wedge}{*} \vec{U} = (T_2 * U_3 - T_3 * U_2, \ldots) \tag{66}$$

Thus with a usual field $\vec{B}(x)$:

$$\begin{aligned} \vec{A} \overset{*}{\cdot} \vec{B} &= \int \vec{A}(x') \cdot \vec{B}(x - x')\, dx' \\ \vec{A} \underset{\wedge}{*} \vec{B} &= \int \vec{A}(x') \wedge \vec{B}(x - x')\, dx' \end{aligned} \tag{67}$$

It is possible to prove that:

$$\begin{aligned} \operatorname{div}(\vec{T} * U) &= U * \operatorname{div} \vec{T} = \vec{T} * \operatorname{grad} U \\ \operatorname{curl}(\vec{T} * U) &= U * \operatorname{curl} \vec{T} = \vec{T} * \operatorname{grad} U \end{aligned} \tag{68}$$

Application: Kirchhoff's representation.—Bicomplex field $\vec{C}$ obeying in Ω external to closed, regular S, the homogeneous Helmholtz equation $\Delta\vec{C} + k^2\vec{C} = 0$.

The extension of (50) to the bicomplex fields and the calculation of $\Delta(1_\Omega\vec{C}) * G$ (where G is the elementary solution $-e^{-ikr}/4\pi r$), immediately gives:

$$\begin{aligned} 1_\Omega\vec{C} &= (\partial_n\vec{C}\delta_S) * G + [\partial_n(\vec{C}\delta_S)] * G \\ &= \int_S (\partial_n\vec{C}G - \vec{C}\,\partial_n G)\, dS \end{aligned} \tag{69}$$

the superposition of a single and double layer. It is easy to establish equivalence with Huygens' representation (Chapter 1, equation (189)).

Note: The integrals corresponding to the two representations can be conventionally extended only to a part of the surface S, with boundary C (oriented with respect to the normal at S). These two representations are not equivalent. For example, for the electric fields, it can be proved that:

$$\vec{E}_{\text{Huyg.}} - \vec{E}_{\text{Kirch.}} = \int_C G\vec{E} \wedge d\vec{s}$$

with the second side being 0 for closed S. This is why the physical optics approximation applied to non-closed surfaces can be dangerous.

7: *Convolution equations.*—Various linear equations are in the general form:

$$A * X = B \tag{70}$$

where A, B are given distributions and X is an unknown distribution, the search for which is actually a 'division' of convolution (known to physicists as 'deconvolution').

Two specific cases. (a) $B = 0$: homogeneous equation with solution H (or an infinite number of solutions, or only one, the trivial solution $H = 0$).

X_0 being a particular solution to (70), $X_0 + H$ is a general solution. The triviality is a necessary and sufficient condition for uniqueness of (70).

(b) $B = \delta$: elementary solution E: $A * E = \delta$. Then:

$$X_0 = B * E \tag{71}$$

which certainly exists when $A, B \in \mathscr{E}'$.

Example
Helmholtz equation in $\mathbf{R}^3$:

$$\Delta X + k^2 X = B \tag{72}$$

B having a compact support (bounded sources). Here $A = (\Delta + k^2)\delta$.

The homogeneous equation is that of the monochromatic waves. It is obeyed in particular by plane waves. The aim of the radiation conditions is to reduce the problem to the trivial case. There is, in addition, the elementary solution $G(r) = -e^{-ikr}/4\pi r$. Hence, for a single solution, the retarded potential:

$$X = B * G \tag{73}$$

which, in the sense of distributions, is indefinitely differentiable: $D^p X = B * D^p G = D^p B * G$.

Convolution systems. $\sum_{j=1}^{n} A_{ij} * B_j = B_i$ $(i = 1, \ldots, n)$ can be represented by (70) in matrix notation:

$$(A)_{n \times n} * (X)_{n \times 1} = (B)_{n \times 1}$$

An elementary matrix solution $(E)_{n \times n} = (E_{ij})$ such that

$$(A)_{n \times n} * (E)_{n \times n} = \delta(I)_{n \times n} \tag{74}$$

allows the column solution to be found:

$$(X)_{n \times 1} = (E)_{n \times n} * (B)_{n \times 1} \tag{75}$$

(h) Fourier Transformation

1: Definition.—The usual Fourier Transform in $\mathscr{S}$ is:

$$\mathscr{F}\{\varphi(x)\} = \hat{\varphi}(y) = \int_{\mathbf{R}^n} \varphi(x)\, e^{-2\pi i(x, y)}\, dx \tag{76}$$

where $(x, y) = \sum_{i=1}^{n} x_i y_i$.

It can be inverted:

$$\varphi(x) = \mathscr{F}^{-1}\{\hat{\varphi}(y)\} = \mathscr{F}_c\{\hat{\varphi}(y)\} = \int_{\mathbf{R}^n} \hat{\varphi}(y)\, e^{2\pi i(x,\,y)}\, dy \tag{77}$$

The Fourier Transform in S′ (the so-called 'tempered' distribution space), $\mathscr{F}\{T\} = \hat{T}$, is defined by:

$$\hat{T}(\varphi) = T(\hat{\varphi}) \tag{78}$$

The conjugated transform $\mathscr{F}_c\{\hat{T}\} = T_c$ is defined by:

$$\hat{T}_c(\varphi) = T\{\mathscr{F}_c\varphi\}$$

It is identified with the inverse Fourier Transform: $T = \mathscr{F}_c\{\hat{T}\}$. If, in addition to T has a compact support, $\hat{T}$ is an analytical function of y with the explicit expression:

$$\hat{T}(y) = \langle T_x, e^{-2\pi i(x,\,y)}\rangle \tag{79}$$

2: Examples.

T_x	$\hat{T}_y$	
δ	1	
$\delta_{(a)}$	$e^{-2\pi i(a,\,y)}$	
$D^p\delta$	$(2\pi i)^{\lvert p\rvert} y_1^{p_1} \ldots y_n^{p_n} = (2\pi i)^{\lvert p\rvert} y^p$	
$\Delta\delta$	$-4\pi^2(y_1^2 + \cdots + y_n^2) = -4\pi^2 s^2$	(1)
1	δ	
$e^{2\pi i(b\cdot x)}$	$\delta_{(b)}$	
$x^p = x_1^{p_1} \ldots x_n^{p_n}$	$(-2\pi i)^{-\lvert p\rvert} D^p\delta$	
$r^2 = x_1^2 + \cdots + x_n^2$	$(-\frac{1}{4}\pi^2)\,\Delta\delta$	

Without usual equivalents, a polynomial and a (multiple) Fourier series also have Fourier Transforms:

$$\left(\sum_p a_p x^p\right)^{\wedge} = \sum_p (-\tfrac{1}{2}\pi i)^{\lvert p\rvert} a_p D^p \delta$$

$$\left(\sum_p a_p\, e^{2\pi i(p\cdot x)}\right)^{\wedge} = \sum_p a_p \delta_{(p)}$$

3: Fundamental property.

$$(T * U)^{\wedge} = \hat{T}\hat{U} \tag{80}$$

$$(TU)^{\wedge} = \hat{T} * \hat{U} \tag{81}$$

provided that the convolutions exist and that in the ordinary products one of the factors can be reduced to an indefinity differentiable function.

Then from above:

$\tau_a T$	$e^{-2\pi i(a\cdot y)}\hat{T}$	
$D^p T$	$(2\pi i)^{\lvert p\rvert} y^p \hat{T}$	(II)
$e^{2\pi i(b\cdot x)} T$	$\tau_b \hat{T}$	
$x^p T$	$(-\frac{1}{2}\pi i)^{\lvert p\rvert} D^p \hat{T}$	

The first formula gives the radiation of the regular n-dimensional arrays, the third (called modulation) applies to the scanning of the radiated beams.

Note

$$(\Delta T)\hat{} = -4\pi^2 s^2 \hat{T} \tag{82}$$

4: Continuity.—$T_k \to T$ in $\delta \Rightarrow \hat{T}_k \to \hat{T}$ in δ'. With the series $\{\psi_k\}$ from page 210, $\psi_k \to \delta$, $\hat{\psi}_k \to 1$ and

$$\mathscr{F}^{-1}\{\hat{T} * \psi_k\} = \int \langle \hat{T}_y, \psi_k(x-y)\rangle\, e^{2\pi i(x\cdot y)}\, dy \tag{83}$$

converges towards T.

Index